DEAD MEMORIES

DEAD MEMORIES

ROBERT CRIGGER

ACKNOWLEDGEMENTS

I thank my parents for doing everything right and also for nurturing my creativity. I was really lucky to have such caring and intelligent parents. I was also lucky to grow up at a time when people still did a lot of things with their hands. As I became older, I marveled at the things they taught me. My father was always the one who helped me with my term papers as I grew up and this was just one big term paper to him. He read my work as I wrote it and gave me helpful pointers.

I also thank my friends for helping me and believing in me. My good online friend Dan Cunningham gave me some very good advice. My friend, Paul Vallejos and his wife Mindy critiqued it for me. For a long time, Mindy was my toughest critic. Toward the end, she seemed to be happy with it too. That was when I knew I was ready to publish. A dear friend read it as I wrote it and gave me feedback daily. She told me not to leave her hanging so I didn't. Her name is Kay Appling and she works with me at the factory. Another friend who works with me at the factory helped me catch several cases of OOPS before they could go any further. His name is Jay Tsinnijinne. I would also like to thank my best friend Bill Jones who took on personal risks to help me do this. I thank everyone who helped me get through this.

CHAPTER 1

Too Good to be True

Steve couldn't sleep. His wife, Rose, lie next to him snoring smoothly and calmly. The sound wasn't a rude earth-shaking tremor. Her snoring was quiet enough that it wouldn't really disturb anyone in another room. The sound was more like a steady murmur of someone talking in their sleep. This sound interrupted his train of thought, which otherwise might have helped him drift into a peaceful lull. He rolled over to find a more comfortable position and he flipped his pillow over to get to the cool side underneath. His mind was slowing down from his day's activities, but he was still blazing with thoughts of the next day.

Everything was quiet in his two-story house. This house was in something that most people would call a good neighborhood. The good neighbors on the right moved in after the crack dealers were dragged away in handcuffs and leg irons. The quiet neighbors on the left kept to themselves because they didn't like anyone. This quiet place was even quieter at 11:15 at night. Steve knew the time because he saw the alarm clock.

A muffled sound of glass breaking downstairs interrupted his peace. In less than a second, he was alert. His heart raced at a breakneck pace, and he was aware of the sound of his own breathing. He heard a sliding sound downstairs. He pulled his arm from under his wife's neck gently and carefully. He slid to the edge of the bed and opened the nightstand. His thirty-eight pistol was there. He opened the revolver to check it for ammunition. It was full. Of course it was full. There was no reason why it would have been empty. He bought it years ago and had not used it much since. He didn't know much about firearms. He tried it out in a quiet wooded area just to be sure it worked. He did this right after he bought it. He got a blister between his thumb and forefinger. It was right in the crotch of his hand. That was a day to remember.

He wrapped a piece of bedding around the weapon to quiet the metal to metal sound as he closed it again. He was catlike as he quietly moved from the bed and across the room. He donned a house coat at the bedroom door, and found a suitable perch to set his ambush. His heart thundered in his ears. He was dizzy with fear and excitement.

His target appeared in his field of fire. It was a man, dimly lit by the light of a street light outside, pouring in through heavy curtains. He could tell by the stiff and deliberate gait, that it was definitely a man. Seconds became hours in his mind. When the man looked up at the head of the stairs, he must have seen Steve, because he made a sudden move. Steve's instincts took over. He didn't think at all. Had he even tried to think, he would have done the wrong thing. He would have pulled the trigger and the pistol would have pointed downward. He did what he was supposed to do and squeezed the trigger, not anticipating the moment that the weapon would discharge its round. At that moment, his thought became a straight line through his arm, the pistol, and finally through the target.

The thunder and flash of the gunfire echoed all around him, but it barely registered in Steve's mind. The sound of his heartbeat filled his mind again. Though he couldn't see the point of the round's impact, he saw by the movement of the dark shape, that he had hit his target. The shape of a man seemed to recoil around a spot in the middle of its chest. The movement of the target became clumsy and unnatural, as it began to tumble down the stairs like a bag of dirty laundry. Steve rose and followed the strangely moving shape, discharging his weapon, until it was empty and the only sound was a click at the bottom of the stairs.

This kind of exciting moment would be unusual for anybody, but for Steve it was almost unthinkable. For him, an exciting day would have had something to do with funding or finishing a big project. He was boring, maybe arrogant, not very charming, and a bit strange. Steve's brand of strange was in his refusal to let anyone tell him what to like. He liked ice cream during a blizzard. His preference in cars had nothing to do with what was popular. He insisted on being an individual. His marriage, miserable as most, had a lot of bad moments and just a few good ones. He was happy at his research work and with his car, which was kind of like a research project in its own way. This event changed everything in his life, but the story really began several weeks before this on the day he gave his big presentation at work.

Stephen Warren looked up from the financial section of his newspaper and glanced at his wife, "Honey! What are these shady Wall Street criminals doing with my money?"

Roseanne Warren looked away from the automatic drip coffee maker and responded, "The same thing they always do. They scam inside information, play the ponies, consult with prostitutes, and check their horoscope." She was barely awake as she said it. She was a pretty woman with fine features. Her blonde hair was past her shoulders and she looked a little bit like Kathleen Turner. When she was in the mood to be charming, she could be very charming. She wasn't in the mood right now. Charm didn't work on Steve anymore. There was too much bad history between them. Rose turned to the scrambled eggs in the frying pan and frowned, "Are you going to eat these eggs this time, or will you push them around the plate with your fork?" She turned to him and gave him an angry stare as she said it, "Why do I bother cooking these anyway?"

Steve suddenly felt uncomfortable and answered quickly to avoid a confrontation, "I'm sorry dear! I can take them with me, but I have a big thing to do today. I have to go in front of the money people and beg again." Steve said and shifted in his seat.

Rose glared at him. She didn't yell, but she spoke louder and with a sneer in her voice, "You do this all of the time. You keep finding excuses not to eat my cooking. You are never home to eat my meals. You never notice the things I do around here. Why do I bother anyway?"

Steve stood up and held her close from behind. He whispered in her ear, "Because in four months, we will be in Puerto Vallarta. We can bathe in the sun. We can go parasailing. We can ride in the glass bottom boat." He swayed his hips from side to side against her, "We can even do that thing in the water that you always wanted to do."

She jumped and turned, "Hey! Knock it off! You dirty old man!"

He squeezed her bottom and whispered in her ear, "Dirty old men have needs too."

She laughed as she spoke, "You'd better leave now because my lover will be here soon." She smiled and put her arms around his shoulders.

Steve answered, "Can he compete with this?" He held his hands apart between her arms.

Rose smiled playfully. She held her hand up with her thumb and index finger close together. She giggled and spoke, "No, but he can compete with this!"

Steve pouted and spoke sadly, "Do you mean you lied to me last Saturday night when you called me your stallion?"

Rose slapped him on the shoulder and told him, "Get out of here and go to work! Someone has to make the house payments around here! I wish you'd get a real job practicing medicine instead of being a traveling salesman and chasing research grants all of the time."

Saying anything bad about his work guaranteed an argument. Steve's tone of voice changed, "I wish you'd stop running around with other guys, running up the credit cards, and spreading lies about me. This counselor's idea of a romantic getaway better pay off. I know I could find a woman who could compete with this!" He held his arms in a circle over his head. She moved her arm to slap him and he stopped her hand with his. Steve became patronizing and began to lecture her, "You know better than that! The counselor told you not to resolve an argument with violence." Rose backed away. Steve saw her retreat, and wanted to enjoy his victory. He lectured her without mercy, "Instead of thinking of ways to get the upper hand and win arguments, maybe you could spend your time thinking of ways to be happy with the nice life you have and make it easier for me to give it to you."

Rose held her middle finger up at him and growled, "I guess you are going out to the garage now to be with your real sweetheart!"

Steve was sarcastic, "Are you jealous of my creation?"

Rose screamed, "That horrible rust bucket suits you! I hope you are very happy together!"

Steve quickly ducked into the garage to avoid any flying objects. Last week she nailed him with a cup. He stuck his head back in and teased, "Temper, Temper!" There were no flying objects, so in one step he lunged back into the kitchen, snagged his briefcase, and recoiled into the garage as if pulled by a rope.

The garage was his happy place. Two vehicles sat side by side. The seventy-two Buick Skylark was his. The eightytwo Ford escort was hers. The garage was where Steve showed his strange sense of humor. The Skylark was his own way of thumbing his nose at the whole world. This car was a Frankenstein creation, with more parts from Checker Auto and Napa than Buick. The rusty body was riddled with holes. The interior looked like an airplane cockpit with toggle switches and gages everywhere. The engine compartment was spotless. The engine came from a Buick Special. The Buick Special was a high performance version of the same car. Steve added headers, an HEI ignition system, dual exhaust, and anything else he could think of, to add extra ponies to his baby. Steve called his car "The Beast", because it had a big appetite and he didn't have to be gentle to it. The Escort was hers and he let mechanics do the work on it. There was no love there.

Steve looked like a teenager as he climbed into his creation. He flipped three toggle switches in sequence on his hand made console, then reached under his seat and touched his special hidden start switch. The car started smoothly and quickly in one turn of the starter. He listened carefully and placed his hand on the dashboard to feel it running, because it was so smooth and quiet. He reached up over his head to the visor and pushed the button on the garage door opener remote. As soon as the garage was completely open, he moved the gear selector into "R", paused until he felt it engage, and pushed the accelerator down one quarter of its travel. The car leaped violently out of the garage. The Beast firmly took traction on the garage concrete floor and repainted some old tire marks from many previous mornings. The car left the ground for a brief moment because there was a dip right outside of the garage. Once Steve felt all four wheels back on the ground, he applied the brakes and slowed down. He drove down the street toward work at a conservative twenty miles per hour. Angry neighbors and speeding tickets made him careful about street racing. This two-second ritual was his daily feelgood routine. For two short seconds, a forty-four year old man was sixteen again.

The ride to work was routine, and didn't take long. He lived ten miles away from his job at Fitsimmon's Medical Center, in Aurora, Colorado. Steve found a parking space in the medical staff area. Doctor Michael Owens got out of his car a few spaces closer to the building. He was also a neurosurgeon, but he was a kid. He was only thirty years old. He shouted over to Steve, "Hey Doctor Steve, why are you driving in something that makes you look like a med. student? Trade that eyesore in on a new car. Get a minivan. That's your style."

Steve chuckled, "You haven't looked under the hood have you?"

Mike walked over to him and said, "Okay, show me."

Steve reached for the handle to release the hood and warned Mike, "Be careful, the overdose of Testosterone might put you in a coma." Steve lifted the hood proudly.

Mike was surprised, "Gee Steve, I didn't know you were a teenager. Didn't the owner's manual say Buick Skylark? Why do you have all of this on the inside and rust on the outside?"

Steve proudly boasted, "It's a disguise. Nobody knows about the 455 in there. That poor Buick Special had no idea it would become a Skylark. Do you remember the old Novas? They were six cylinder beaters out of the factory. Teenagers made them into fire-breathers. Novas have a racing reputation, Skylarks don't. The Buick 350 is a big block with lots of torque. Before I sized up, I pulled a 44-passenger school bus up a hill with it. Don't underestimate a Buick V-8. It may not reach 150 mph, but it's a lot of fun getting to sixty."

Mike was confused; "This sort of behavior signals male menopause. You do look pretty old. Do you need your medication adjusted?"

Steve just smiled; "Sure pal. Just go back to your Lexus and dream of what real torque feels like." Steve and Mike parted at the parking lot. Mike looked down and shook his head as he walked away. In a few minutes, Steve was standing in the lobby outside of the conference room. He poured himself a cup of coffee. The secretary greeted him, "Good morning Doctor Warren. The funding board will be ready for you in twenty minutes." She then looked away and returned to answering the phone and working on her computer. Steve busied himself as he waited, by sipping his coffee and brushing up on his presentation. His coffee was black and without sugar. He didn't like pretending to drink coffee by hiding the taste with cream or sugar. Half an hour later, one of his colleagues stepped out of the boardroom. The man looked sweaty and nervous. He glanced at Steve and said, "Your turn Steve." The secretary looked up too and spoke up, "The board is ready for you now, Doctor Warren." Steve straightened his tie and marched confidently into the boardroom.

Doctor David Anderson was seated at the far end of the table. He stopped practicing medicine and began running things fifteen years ago. He was now the president of the medical company, which ran the hospital. Doctor Anderson smiled at Steve, "Doctor Warren, what do you have for us this morning?" Steve placed his briefcase on the table, then pulled several portfolios out and passed them around the room.

Steve addressed everyone in the room, "Gentlemen, what I have for you is too good to be true. The folders I have passed around to you list the interested parties, our prospective customers. The customers are colleges, research programs, and medical foundations. I have bites in other industries as well. Several electronic manufacturers made big offers to shut out their competition. The investment cost is seventeen point four million dollars. The return on the investment is ninety-two point five million dollars and climbing. Gentlemen of the board, this is a cash cow."

A local bank president looked up from one of the portfolios and held his finger on a spot on the page, "Doctor Warren, I understand you are a neurosurgeon, but your customers are sociologists, psychologists, and mental health hospitals. I see a foundation for Alzheimer research.

Down here I even see five million from the CIA. What does a neurosurgeon have to offer these parties?"

Steve had a big smile on his face, "Sir, something nobody else can give them. I am offering a complete understanding of the human mind. I intend to build a computer model of the human brain, complete with the software to translate it into something a computer can understand. Accomplishing this task would allow us to read every thought in a man's memory. We could see everything he has ever seen. We could hear everything he has heard. We could find out what happens in the subconscious." Steve walked over to the board member, crouched behind his seat, and looked him in the eyes, "Have you ever seen a thought? What does a man see after he dies? What is it like to be born?" Steve stood up and walked back to the end of the table.

The bank president stood up and glared at Steve, "This sounds like a lovely traveling show. Were you going to part the Red Sea or ride a magic carpet too? Why do you think you can do this?"

Steve smiled confidently at the board member, "Sir, it's so simple, it's almost impossible." He reached into his briefcase and pulled out a dry erase marker. He walked over to the board and proceeded to draw a picture of a human brain, and another of a desktop computer. He spoke as he looked around the room, "Gentlemen, the human brain has about 100 billion brain cells. In computer terms, that is 100 Giga-bytes. We have desktop computers smarter than that." He pointed at the computer he drew; "We know everything there is to know about that." He pointed back at the brain, "That is just wired differently. To completely map it out and trace every connection would be a monstrous task. That is why nobody has ever done it before. We know which parts of the brain perform which functions. We know where feelings happen. We know where pain is felt. We know which parts of the brain move muscles. All of this we know, yet we have never figured out what a thought looks like. Think of the brain as a big cluster of wires and switches. To really understand the way it works, we must follow the wires and switch logic. That is the hard part. We have microscopes that can make pictures and computers that can read them. The machines used to make slides were around when I was a kid. Ten years ago, it would have been unthinkable! Now it is just creative and tedious." He held up a stack of folders from his briefcase, "I have resumes from computer engineers, programmers, surgeons, and a couple of mathematicians. The cost I quoted you buys a staff of fifty-three special people at forty hours per week for six months. This cost includes the equipment, space in an office building, and my salary of four point three million dollars for delivering this handsome child to you before anyone else comes up with it."

Doctor Anderson, the president spoke up, "That's a handsome salary Doctor. Aren't you shooting a bit high?"

Steve looked directly at him, "Sir, you'll have that amount with the first sale. Could anyone else deliver this to you?"

Dave Anderson smiled, "You have delivered before. You must be planning your retirement." He looked around the room for approval. Everyone was smiling and nodding, "Doctor Stephen Warren, we will call you in about two weeks. We will discuss this further. Thank you for your presentation. You will hear from us."

Steve felt good about himself. He gathered up the papers he needed and left the room. He overheard one of the board members whisper, "Last time we made fifteen million on his work. He only sounds confident when he has something good." Steve strutted out to the parking lot and climbed inside the Beast. He flipped his switches and started his car. He could feel the engine churning. Every time he lightly pressed on the accelerator, the body of the car twisted a little bit. He spoke under his breath, "Yes, I love you too Baby." Steve paused for a moment and decided not to go home right away. It was time to play.

Steve turned East instead of going North. He knew of a special piece of road, which he liked to play on. This special place was miles away from the city and surrounded by fields in every direction. Much of the road was gravel. Some of it was bumpy. There were deep ditches on both sides, and the road was barely wide enough to be used as two lanes.

Steve arrived at his special personal racetrack in half an hour. He looked around to see if anyone was watching. The coast was clear and it was time to play. He came to a complete stop and breathed deeply. He glanced at all of his gauges to make sure that everything was performing at peak. He felt his heartbeat as his excitement grew. He punched the gas and the body of the car lurched forward roughly. The tires lost traction, but the body surged forward at an alarming rate. He could feel the force of the acceleration as he was sucked backward into his seat. His heart pounded harder. The speedometer needle looked like the second hand of a stop watch. The RPM gauge reached three thousand. The engine wasn't even working hard. At eighty miles per hour, the road became so rough, that the car bounced to the left and right. Steering was hard to do because the front tires weren't on the ground enough to direct the car anymore. He narrowly missed the ditch and felt the grass under his tires. Steve backed off and lifted his foot from the accelerator pedal. He slightly over steered and felt a hard twist as the car touched the ground again. The feeling in Steve's stomach bordered fear and exhilaration. The car slowed to twenty-five miles per hour, and Steve put his foot on the accelerator again. He sharply turned the wheel left and goosed the gas. The rear of the car swung around wildly. Once the front of the car was almost pointed in the opposite direction, he released the accelerator pedal and turned the wheel to the right. The car snapped back quickly and headed down the road in the opposite direction. Steve felt good after his brief celebration and went home to relax.

Friday, two weeks later, the phone rang. Steve answered it. It was Dave, "Congratulations, you have the funding. The money will clear in two weeks. We found some of your equipment and gave you office space. Some of what you needed was in our warehouse. We spoke to a few specialists in the field and they confirmed what you said. They didn't want to do it, even at your price. We did a little fishing and found a few more customers. I found one international customer willing to pay eight million for the product and they only want half of the brain. IBM promised fifteen million for the whole product. It looks like a rich harvest. Your cost looks like it will come in less than fifteen point one. I want you here on Wednesday morning in three weeks, at seven in the morning. I need the team here by the following week. Do you have the brain yet?"

Steve was containing his excitement, "Not yet Sir. I have been scouring the morgues. I need a body, which has no family, and has been donated to research. Even then, I have to fight with red tape. I almost had one last week. Four thousand in the funding has been put aside for burial expenses."

Dave laughed, "I'm glad to see you have been working on that. You have thought of everything haven't you? I want daily meetings on this. If this pays off like I think it will, your retirement might be better than you asked for. Steve, make this work out. I am counting on you."

Steve was trembling with excitement, "Thank you Sir. I will be there."

CHAPTER 2

THE BRAIN

S teve spent most of his first day on the project in meetings and on the phone. Before that day was over, he had two team members on salary. The two members began setting up the computer equipment and operating equipment. Jacqueline, or Jackie Sears was laid off from an electronic manufacturer North of Denver, and her severance package was getting skinny. She arrived at work in two hours. Jackie was single, thirty-two, and very disillusioned with her line of work. She had a slight potbelly, but a very pleasing face, and even her glasses looked good on her. She had a certain Sandra Bullock about her which gave her a kind of warm cozy glow. Jackie was just nice to be around. She wanted to try something else and she wanted a position with more responsibility.

Michael Owens was already on the staff at Fitsimmons. He was a very good neurosurgeon, but itched for a chance to get in on the new action in his field. Mike had a good frame and good looks. He handed Steve a resume at lunch one day and informally interviewed with him for the chance to get in on research. He was smart, young, and full of himself. His task was to remove and handle the brain. He cleared his schedule and reported to the project in twenty minutes. Both people wanted this project very badly.

Steve looked at his reflection in the mirror before he left for home, "Yes, Ladies and gentlemen of the science community, I am proud to accept the Nobel Prize for my lifetime achievement." He grinned as he spoke, "You did it, you slick son of a gun. You made the big leagues." Steve was pleased with his reflection. He was mostly thin, but had a potbelly. He weighed about one hundred eighty pounds. He looked a little like an accountant, but had some muscle tone. He gathered his hat and coat as he walked out of the door.

Steve stopped for pizza on the way home. Larry Curtiss was behind the counter and ready to take his order, "Good evening Mister Warren, will it be a wild garlic pizza again?" Larry was about six feet tall and weighed about two hundred pounds. He had a few acne scars on his face, but could easily disappear in a crowd. Larry was new in town and had just gotten this job and a place to stay.

Steve replied, "Yes it will! I have something to celebrate.

How is this job working out for you?"

Larry happily replied, "I have a roof over my head and a job. I'm better off than I was a week ago. Who could ask for more, unless it were a reason to celebrate like you have."

Steve's excitement showed in his voice, "I just became a really big deal. I could be talking to Oprah next week. National Enquirer will want to know what I had for breakfast."

Larry grinned, "I know the answer to that one. It will be cold wild garlic pizza."

Steve laughed, "That's right! National Enquirer will pay a fortune for that scoop."

Larry stepped away from the counter; "I'd better hurry up and make this thing then. I have to retire on something!"

Steve asked, "What kind of retirement plan do you have now?"

Larry grinned, "A bottle of whiskey, newspapers for a blanket, and whatever I can steal from a Dumpster. I'm still working on it, but I'm definitely going to fire my financial advisor tomorrow."

Steve liked the joke, but became serious, "Good idea, I need to fire mine too." He went over to a bench seat, watched the news, and waited for his pizza. After a short time, Larry alerted Steve that his pizza was ready. Steve collected his food and left. In twenty minutes, he was in his garage, and carrying the food into the house, "Honey, your feast is here!"

Rose answered from another room, "Just a minute." A moment later she was in the kitchen with Steve, "Let me guess, pizza. Why did I cook lasagna?"

Steve answered, "So I could take it to work with me. Thank you for making my lunch."

Rose growled, "I should make you into lunch."

Steve flexed his left arm and pointed at his bicep muscle, "That would feed a whole family." He grabbed his thigh; "This would wipe out hunger in Iraq." He held the pectoral muscle on his chest; "Beef anyone?"

Rose stopped him, "Okay, enough already! Don't turn this into a biology lesson. Where's the pizza? Is it wild garlic?" Steve smiled and opened the box. She took a piece and tasted it; "It's better than last time. I won't make you into lunch this time."

Steve laughed, "Good, I was worried. You know where I sleep and I've seen you dress a turkey. I don't want stuffing there."

Rose choked on her food and held two fingers over her mouth to keep food from falling out as she spoke, "Knock it off! I don't want to imagine what that looks like while I'm eating. Keep your medical stuff at work. You seem more excited about this job than the others. How much are you getting?"

Steve lied, "The same thing I always get. I just like the new challenge. We are doing okay and the house is still getting paid for." He thought in the back of his mind, "Enough money to get a good divorce lawyer and send you packing," He spoke out loud, "You know I love you."

Her expression changed, "Love? That's what you call it? I'm alone all of the time. When you are around, you don't talk to me. When was our last romantic evening? When was the last time you gave me flowers? When was the last time you called me from work to let me know you care? I don't know how we are doing because you don't tell me." She punched him in the shoulder and began to cry.

Steve was apologetic, "Honey, you know I care. I have to work hard to keep everything going. I still take care of you. I give you a nice big house to be lonely in." He held her close and squeezed her bottom.

She held him too and whispered, "You do know you are an arrogant patronizing jerk, don't you?"

Steve replied, "Yes, but that's what you used to like about me."

She pinched his back and whispered a little louder, "I never liked that about you."

He answered, "Ouch! Okay, what did you like about me?" He waited, but heard nothing. He asked, "Well?"

She said, "Shut up! I'm thinking!" She still held him, but backed away far enough to see his face. She asked softly, "What did you like about me?"

Steve paused and smiled, "You were always pretty. When I first noticed you, I loved the adorable expressions you made. When you brushed past me in the cafeteria that time, I tingled all over. Everything about you was soft, sweet, and feminine."

She smiled and blushed. She whispered, "You were so cute and macho with your car. It was so hot with the nice paint and nice interior. It was fast too. You took me on some scary rides. My father hated you. That clinched the deal. You were smart too. I knew you would get somewhere. I felt warm and safe around you."

Steve softly brushed the hair out of his wife's face and kissed her. First he kissed the upper lip, then the lower. As he kissed her, he caressed her face.

Rose smiled and giggled, "Are you ready to prove that?" Steve took her hand and turned it so her palm was facing his face. He then gently licked her palm. When he wanted to, he had a touch softer than a feather. He didn't even leave any moisture on her hand. He saw the goose bumps rise on her arm. She gasped, "I guess so! The bedroom is that way." She tugged on his collar.

Both of them walked hand in hand to the stairs to go to the bedroom. Steve paused briefly to turn off the lights. For the next hour, giggling and happy sounds filled the house. Rose fell asleep quickly. Steve lay there and thought about work. He was like a kid on Christmas Eve.

A few hours later, Steve heard a muffled sound of glass breaking. Steve was suddenly very alert. He gently and quietly removed his arm from under his wife's head. He had an uninvited guest. This guest was quiet and didn't turn on lights. Steve thought, this is a thief or worse. He wanted to have the element of surprise and every other advantage he could get. He quietly and deliberately slid across the bed to the nightstand and pulled the drawer open. As he opened the drawer, he lifted it slightly so it wouldn't make a scraping sound. Steve heard footsteps on the carpet. His heartbeats pounded in his ears. Steve pulled a thirty-eight revolver from the drawer. He wrapped the bed sheet around it to muffle the sound. In the dim light from the streetlights outside, he could see that the pistol was full of ammunition. He slowly, deliberately and quietly squeezed the pistol shut with the sheet around it to make it quiet. The click thundered in his ears, but it barely made a sound. Steve moved like a cat toward the bedroom door, carefully placing his feet so that they made very little noise. He heard silverware rattle in the kitchen. Steve was shaking like a leaf. The house was cold, but he was also terrified. He pulled a housecoat from the door, wrapped it around himself, and moved to the top of the stairs. He stopped there. Since the sounds were moving closer, Steve was sure that whatever was in his house was coming to get him. He decided to wait there. Steve lay on the floor at the top of the stairs to make his profile low and small. The pistol was in his outstretched arms and pointed in the direction he expected his intruder to appear in.

Steve was trembling and his heart pounded in his ears. He heard another sound. It was closer than before. Steve was extremely alert. Every second seemed to last an hour. His mouth was dry and had a stale taste in it. His stomach felt heavy, and seemed to be moving toward his throat. He was uncomfortable and shifted to keep from cramping. A fuzzy figure moved into his field of fire. Steve slowly and deliberately followed the figure with his firearm, aiming for center of mass. The figure moved suddenly as if to fire upon Steve, and Steve squeezed the trigger. He didn't pull the trigger as if in a panic. He almost instinctively let the weapon discharge. The thunder of the weapon filled the house. In the muzzle flash, Steve could almost make out a face. The noise was deafening, but he still heard his heartbeat in his ears. He released the trigger and the click of the trigger spring returning echoed. Steve suddenly felt a rush of strength and energy. He lifted his body slightly up off of the floor and keeping the pistol aimed at the center of mass of his target, he squeezed the trigger again. The hazy and shadowy figure began to fall away from him. The movement was not natural anymore. The shadow tumbled clumsily away and to the floor as gravity took over. Steve moved as if in pursuit toward the retreating shape. He met his target at the bottom of the stairs and pulled the trigger until the pistol stopped firing. Steve was still trembling, but now more with excitement than fear. He looked around to see if there was any other threat near him and he leaped away from the mess at the bottom of the stairs. He moved toward a hazy spot where he knew a light switch was. He fumbled around until he felt the switch and lifted it. The brightness hurt his eyes for a moment.

It was Larry. Why was it Larry? The body was in an unnatural pose crumpled up at the bottom of the stairs. Eyes stared straight up. Five red and black marks walked across his chest and a dark red puddle covered the floor. It didn't move at all. Steve poked it once on the face and the head bounced right back where it was a moment ago. The mouth was wide open. A line of blood went from his nostril and disappeared into his hair. Steve recoiled from this ghastly scene.

Steve was still shaking and felt his stomach going up into his throat. He put his hand over his mouth to stop it. Steve ran to the downstairs bathroom and choked on what was coming up his throat. Beads of sweat ran down his face as he sat on the floor with his chin hanging over the toilet bowl. He choked and his body tensed up. Noise upstairs meant that his wife left the bedroom and was looking to see what happened. She called to him then screamed and he heard a door slam. Steve couldn't talk because he was busy doing something else. In a few minutes, his stomach was empty. He brushed his teeth and rinsed his mouth with mouthwash. His throat still burned and his stomach ached.

Steve left the bathroom and walked over to the phone next to the couch. He sat down and picked up the phone. He was too shaken up to look in the phone book for the right number for the police, so he called 911. He couldn't look at the body at the stairs. He closed his eyes. The person on the other end of the line began asking questions. Steve could only say, "I shot a man. I have a body. I shot." He babbled and stammered. Steve couldn't speak. The part of his brain, which controlled speech, was out of order, while other parts of his brain galloped ahead at a breakneck pace. The shooting event was playing back over and over again in his mind. His heart thumped in his ears and felt as if it were in his stomach. Every few moments, the voice on the other side of the phone said something, which woke him up again, and he responded with, "Huh?" or "I". The voice on the other end of the line stopped and there was a knock on the front door. Steve stumbled to the door and opened it. It was the police.

Steve didn't say a word. He backed away from the door and left it wide open. His mouth moved, but he didn't talk. He backed over to the couch, sat down again, and picked up the phone which was blaring a busy signal into his ear. The police didn't wait in the doorway for long. Steve didn't respond to any of their questions, but they could see the body on the floor. One officer entered with his weapon before him, found Steve, and kept it pointed at him. Steve stared into blank space in front of himself with his mouth wide open, and didn't move. Another policeman followed, took control of Steve, searched him for weapons, and revived him with smelling salts. The officers turned him around so he wouldn't be looking at the body.

Steve was more responsive now. He began answering questions. In a few minutes, the house was full of men in lab coats. Steve and Rose were each talking to police officers in their front yard. Morgue personnel carried the body out to a van and left. Steve and Rose left in separate police cars and went to the police station. Both of them were questioned in separate rooms. Rose was dismissed from the station at four fifteen in the morning. A police car returned her to her front yard. She wasn't even a witness, because she was asleep when it started. Steve was dismissed and returned home at five forty five. Detective Edward Morris let Steve get home so he could go to work. Steve's reputation, cooperation, and the evidence convinced the detective that Steve was not an escape risk. Everything suggested that he acted in self-defense. They had no reason to keep him.

When Steve got home, he felt drained. He had answered every kind of question. He had told them about his work and project. He told them about his marital problems. He even told them about the pizza he bought from Larry.

Steve had enough time to bathe, get dressed and go to work. Steve arrived at his office door ten minutes early. Dave met him at his door, "Doctor, you have a large staff in your lab area. Twenty-two people need direction. I also have crews ready to install equipment and remodel. I'll bet you have a few miracles to perform today. Do we have a brain yet?"

Steve was slightly agitated, "No Sir, we do not. Sir, I had a very big problem at home last night. I shot an intruder. I didn't get any sleep at all. I am still pretty shaken up. Can we down shift for just a moment?"

Dave frowned, "I understand why you might have a problem right now, but you still have a job to do. This is your project and the clock is ticking. I am counting on you. I can buy some time with the people you brought in yesterday. I need you to collect yourself by noon. I don't think you understand the situation. This is not a factory. You don't have sick days. You can't take vacations. Thirty investors are on the phone every day and they don't accept excuses. I have to answer to these people every day, and you have to answer to me! This is not a staff position. You bought a project and this is the price. If I could replace you with another surgeon, I would, but it has to be you. Make it happen!"

Steve held his anger back, "Yes Sir. I will be productive and efficient by then."

Dave nodded and walked away.

Steve pulled his office door keys out of his pocket, unlocked his door, and walked in. Before he got to his desk, the phone began ringing. It was the hospital morgue, "Doctor Warren, the body you asked for is ready for you to pick up. The person you shot last night is available to you; no next of kin, no arrangements. The police dropped him off for you and said you can have him."

Steve was puzzled, "Bill, I don't understand. This usually takes weeks. How did it happen so fast?"

The man on duty in the morgue replied, "The man was a drifter. His name was an alias. The real Larry Curtiss died five years ago in New Jersey. The police forensics lab people collected all of the evidence they could. They know about your project and want to collect evidence from your results. You will get a call from them today. You have been drafted for the forensic team."

Steve ended the conversation. The next call was from Edward Morris, "I gave you the body. I am interested in your project and want information from that man's brain. I want you to help me question a dead man. Let's just say that you are an unofficial extension of the forensic department. This will help me clear your name too."

Steve was nervous, "Yes Detective, I will cooperate fully. It will take a few weeks before we have anything at all. I will give you whatever you need."

Edward Morris ended the conversation, "Thank you Doctor. I will be in touch. You'll hear from me soon if I don't hear from you."

CHAPTER 3

THE PROJECT

Steve got up from his chair and walked to his newly assigned research laboratory. A crew was removing walls. Another team of men rolled computer stations into the room. The research team members took the stations and hooked them up as soon as they arrived. The activity was loud and confusing to Steve. Telephone crews were setting up phone lines and a modem for one computer terminal. At the other end of the space, an operating room took shape. Near the new operating room was some equipment under a canvas tarp. The tarp and equipment looked very old. Steve stood in one place and looked around. He saw Michael Owens approach.

Michael looked impatient, "Steve, I'm glad you are here. We are ready to go to work if you have our donor." Michael didn't see a reaction on Steve's face, "The brain donor?"

Steve didn't react for a second, then he grinned, "That's because you haven't gone to the morgue to pick him up yet. Just say you are there to get my body. Can you find two surgeons to assist? Get busy on the extraction right away. I need optic nerves, eyeballs, and auditory sensory organs still attached. We will follow them to the brain and see how the information is processed."

Michael replied quickly, "Right away! Will you be in on the extraction?"

Steve answered, "Sounds nice, but I have to do something else. It's just a dissection anyway. You don't need me." Michael left the room to get the body. Steve turned and addressed the whole room, "Who is in charge of the nitrogen freezing and slide preparation?"

"I guess I am!" came from across the room. A tall thin man in his sixties stood up, "I came back from retirement to do this."

Steve smiled, "Doctor James Jordan! I am glad you could make it. We are going to be making long slides and a lot of them. The computer scanners and software will build a model from it. Your baby is here and ready for you to operate it."

James walked over to Steve. Steve pointed at the tarp. James lifted the cover and grinned, "Old faithful. I remember that one. It's an antique all right. See where I marked it?" Several of the computer operators chuckled. Jim looked at them and laughed, "Without this machine, none of that fancy stuff means a thing. Without this, you don't have any slides to scan into your computers." The chuckling stopped.

The slide making method involved soaking the subject in a warm liquid, then freezing it. The tiny cell structures would then be rigid enough to slice with a blade. If the sample were not frozen, then the cell walls would give way and the blade would just make a mess.

Steve asked Jim, "Why does a Family Practice Doctor know how to do this?"

Jim smiled, "I wasn't always a Doctor. I worked my way through Medical school as a lab assistant."

Steve turned around and saw Michael bring the body into the space. Steve walked over to Michael and whispered to him, "I put that man there. He broke into my house and I killed him. I have a little bit of a problem with that right now. I didn't get any sleep last night because I was at the police station. I will be leaving soon. I have to get some sleep. I want you to handle it. You know what to do." Mike nodded and turned away to begin the operation. Steve looked for Jackie and called her over, "Jackie, finish setting up the computer stations. You are in charge of the computer team. You know what to do. Just get everything to work. I have to be somewhere else for the rest of the day." Jackie nodded and returned to working on the computers. Steve left the room and went to Dave's office, "Sir, the brain is being extracted right now. All of the equipment is being installed and hooked up. The man who operates the blade is getting his machine ready. I left Jackie Sears in charge of the computer people. Until the brain is out and frozen; we can't do any more. I need to go home and get some sleep."

Dave answered, "Okay Doctor, you can go home. I need you to be on board tomorrow."

Steve nodded and walked out of the office. He went to sleep as soon as he got home and didn't wake up until he had to get ready for work. The next day was Friday. Steve arrived an hour early for work because he was the project manager and it was a good management practice to be earlier than the troops. Showing up early sent a message of professionalism and enthusiasm. The team needed to know that Steve cared and that this undertaking was important. Showing up late would have had a disastrous effect on the operation.

Jim greeted Steve at the door, "Finally showed up huh?"

Steve was disappointed that he wasn't the first one there, "Good morning to you. How long have you been here?"

Jim smiled, "Just half an hour. I got up three hours ago and couldn't sleep. I was thinking about the project and the best way to handle the subject. I have never made slides this big before. This is a challenge." Jim waited for Steve to open the door, and then he led Steve to the refrigerator cabinet where the body was stored. He pulled the drawer open. The object there didn't look human anymore. The face was gone and the eyeballs were draped over a gooey mess, "Your boys are taking a long time because they decided to remove the bone from the brain instead of the brain from the bone."

Steve was impressed; "Doctor Owens is doing the right thing. The optic nerves will be fine and the ears will too. It takes a while, but this is the best way to do it. I am glad I hired him. Is your equipment ready too? Is the tank at temperature?"

Jim replied, "I'm all set. I had to make a carrier big enough for the subject. I am going to give the subject twelve hours to soak the fluid in and another day to freeze it. This project isn't exactly by the numbers, but I believe it will work out fine. I made bigger blades for the machine too. I have never seen it done like this before, but I feel good about it. If you want to save some time, you can let me in the lab tomorrow and I can put it over to the tank. By early Sunday, I can slice it if you get me in here again. You may want to bring a couple of extra folks in to help get slides ready. We can only cut it for three hours before it has to go back into the tank again."

Steve smiled, "Yes! Yes, you are right. That will keep us from having fifty people standing around and waiting for the slides. I'll open it for you." He pulled a card out of his pocket, "If you need to get in, call me. I only live ten miles away."

Jim grinned, "I could have a few hundred slides done. That'll give these hot shots something to play with. Those knuckleheads will be a week behind by Monday."

Jim and Steve both laughed over their morning coffee. In fifteen minutes, the laboratory began to fill up. The three Surgeons began to cut the subject with small rotary saws. The noise and smell of the cutting operation was too much for the rest of the crew, so a temporary barrier and ventilation hood, were installed to remove the smell and mask the noise. The exhaust hose led right out of the door of the lab and down the emergency stairs. Steve moved continuously between the surgery space and the computer area to keep everything moving forward and call in outside assistance when it was needed. Shortly after noon, Dave Anderson walked into the space and found Steve to get an update, "I can see you are busy. Update me later." He left.

In a few hours, the brain was free of everything around it, and the surgeons called Steve over to tell them where to sever it from the body. Steve pointed at a spot two inches away from the brain stem. Mike sliced it with a scalpel and lifted it up from the table. He turned it upside down and lowered it into the container full of liquid, which Jim provided for them. He draped the eyes and optic nerves back toward the base of the brain. Mike walked the container through the surgical curtains to Jim. Steve was right behind him. Steve tapped on the table where Jim was relaxing and reading the sports page of his newspaper. Jim glanced over, smiled, and stood up. He took the sample container, pushed it closer to the tank, grabbed his coat, and waved to Steve as he left. On his way out, Jim smiled and said, "We can put it in the tank tomorrow."

Steve looked around at the three surgeons who had just finished preparing the brain, "Let's get you paid." The three surgeons removed their masks, gloves, and aprons. Steve picked up the phone and called the president, "Sir, the brain is in the container and soaking. The surgeons are ready to be paid. I'll send them to you." He paused for a moment and replied, "Yes Sir." Steve put the phone down and turned to the surgeons; "He is ready for you."

Mike spoke up, "Steve, we are happy to get paid, but we want to stay with the project. We want to see the whole show. This is exciting stuff. We all got into this field to study the brain, and this is like the next page. We want to stay."

Steve smiled and spoke, "I know what you mean. Tell the president that. You can work the terminals if you don't mind taking less pay." The surgeons nodded and left the room. Steve walked over to Jackie and whispered, "Update please?"

Jackie responded out of breath, "Yes! The terminals are up and running. A big screen got plugged in over there at the other end of the room. All of the terminals feed it through that tower over there. The scanners now work with the microscopes. A couple of slides would help. Now all we need is a little help with morale." She pointed at Steve.

Steve nodded, "Yes! You are right." He turned to face the whole room, "Hello team. I haven't been here for you very much up to now because the other part of the project needed attention first." Everyone in the room stopped what they were doing and listened to Steve. He continued, "By Monday, we will have a few hundred slides to scan and study. I will take volunteers now for Sunday to prepare slides. We will begin at about noon. On Monday, we will have plenty of work for everyone to do. I would like for most of your questions to be handled through Jackie." Five volunteers raised their hands right away. Jackie took their names. Steve finished, "Right now, there isn't too much more to do. Jackie will keep a skeleton crew and let everyone else go home. Enjoy the time off while you can." Steve turned to Jackie, "What else can I do to help?"

Jackie replied, "You just did it. Everyone wanted to know what is going on. Two more software engineers and I are working on software to recognize cells and map them into the computer. This software will assign each cell a code number. Without that, the mess would be unmanageable. The hard part is connecting the dots and tracing the neural pathways between cells. I have ideas for software to do that too. I just haven't gotten that far yet."

Steve smiled, "If you can do that, then we won't need fifty people to do this job. The machine could do the work faster than we could." Mike and the other two surgeons walked back in the room. Steve turned to Jackie, "These three men will be working terminals. Get them set up with what they need." Jackie took the surgeons in tow. In a few minutes, most of the operators went home and Jackie brought in three more terminals for the surgeons. A security crew installed thumbprint recognition sensor locks at a few of the doors. Steve and Jackie entered their prints for access. Dave came down to do the same. In two hours, the work for the day was done, and everyone else went home.

Saturday morning, Steve arrived at work at nine thirty. Jim was waiting at the door when he got there, "Finally got up huh?" Steve grinned and put his thumb against the new door lock. Jim pointed at the new hardware on the door, "New toy huh?" The door opened, and both of them walked in. Jim went straight over to the brain, suited up, grabbed two hooks, and placed the brain in the tank. He remarked as he closed the lid to the tank, "This place would be perfect if we had some music."

Steve spoke, "We can listen to radio or CD's on that terminal over there."

Jim smiled and replied, "Good, just don't play any of that rude gang-stah rap. That stuff is shoved down my throat every time I drive somewhere. I swear I'll walk off the job if you do."

Steve grinned and added, "I will too." Both of them left the room and went home. The next day was Sunday. Steve arrived at work at eight fifteen. Jim was waiting at the door and smiling. He held up music CD's for his music choices. Steve said, "If that's Lawrence Welk, I'm walking off of the job." He stepped closer and read the titles out loud, "Eagles, Uriah Heep, Aerosmith, Rolling Stones. Okay, I don't have any excuse to go home and walk around the house in my underwear." Steve unlocked the door with his thumb.

Jim shook his head and mumbled, "Lawrence Welk! What kind of senile nut do you think I am? Underwear? Please."

Steve answered quickly, "Never mind me. I take it back. You can pick all of the music. I was wrong." He mumbled quietly, "I'm surprised I didn't see Iron Maiden."

Jim spoke up, "That's at home." He walked across the room to the tank and put on his protective clothing. He picked up the hooks and pulled the sample out of the tank. Jim removed the container, which was so brittle that it fell apart. He then installed a device, which looked like a clamp to the block of ice. This clamp held the block firmly and connected to the machine with the cutting blade. Jim turned the machine on and the blade moved back and forth against the block of ice, "It will be about forty minutes before we have anything useful out of this thing. I am starting at the back of the brain and working forward. In three hours, we have to put it back in the tank again." The blade peeled its first piece off of the block. It fell to the table. The room filled with the sound of the Eagles. Jim smiled and remarked; "Now that's more like it." He glanced up and saw Steve at a computer terminal holding up the CD cover to the Eagles. The blade moved rhythmically back and forth, keeping beat with the music as it peeled thin slices of the block.

Steve moved around the room turning on computer equipment. The room woke up and became alive as screens became bright. The hum of hard drives joined with the music. Jim placed pieces of paper under the blade to catch the slices. He cleared these slices away and replaced the paper. Jim spoke to Steve, "You'd better get over here with some slides. It won't be long." Steve was there in a couple of minutes with a stack of glass sheets. Jim looked at him; "You'd better get more than that." Steve returned with two more stacks. Jim spoke quickly, "Now." Steve didn't know what to do. Jim took a sheet of glass and put it under the falling slice, "Keep up Doctor Warren." Steve began to place and replace the glass sheets. He had to move quickly to keep up with the machine.

A few minutes later, Dave Anderson walked into the room and shouted, "Hello?"

Steve spoke loudly, "Right here. We need you over here.

Help please!"

Dave walked over and explained, "I wondered why the lab door was open."

Jim said, "Never mind, just help bring sheets of glass and put these in order somewhere. We have to move fast because the sample has to go back in the tank soon."

Dave nodded and followed instructions. He placed sheets of glass all over the room, "I didn't know you were going to be working today. Why are you doing this on a Sunday?"

Steve replied quickly as he hurried to catch another slice, "We didn't want to have fifty people standing around and waiting for something to do. Until now, we couldn't get any slides. This will give everyone something to do on Monday."

Dave picked up another slide, "That's wonderful Doctor, but I can't stay long."

Steve answered, "You won't have to. Volunteers will be here soon. By the way, this is your update on the progress of the project."

Dave was cynical, "Gee, thanks!"

One of the volunteers arrived in the doorway, "Is anyone here?"

Dave yelled to him, "Yes! Over here! I'm glad you showed up."

The volunteer replaced Dave, and Dave vanished. For the next two hours, volunteers arrived and helped make slides. One of the volunteers organized and boxed the slides. Two of the volunteers started scanning the slides into the computer. At eleven thirty, Jim announced, "It has to go back into the tank now! Show is over!" Jim put plastic over the freshly cut area. He put his protective equipment on and put the sample back into the freezing tank. He walked past Jackie as he left the laboratory.

Jackie saw everything going on in the room and walked up to Steve, "I got here early so I wouldn't miss Doctor Jordan's magic show."

Steve answered, "It was a good show. Sorry. You can see it tomorrow."

She just said, "Okay." Jackie looked around the room and began to give orders to the computer operators, "Brian, you can handle that by yourself. Greg, start scanning slides. Brian, I want code numbers on those."

Steve walked up behind Jackie and whispered, "You don't need me anymore. I'll see you tomorrow. Bye." She stood there with her mouth open and watched Steve walk out of the door. Her new boss had a nasty habit of walking away and leaving her alone to run the show.

CHAPTER 4

THE BREAKTHROUGH

In Monday morning, the laboratory was busy and exciting. The new large screen on the other side of the room displayed the part of the computer model,

which was scanned into the computer. The brain cells were in gray. The parts, which had complete neural connections, flashed bright white lights, like lightning bolts, as each connection was completed. Every few seconds, another spider web-like flash appeared on the large screen. At the lower right corner of the screen was a number indicating the percentage completed. The number was two percent. Two percent of the brain cells were coded and connected. The other end of the room was busy too. Jim was slicing the sample and two helpers prepared slides and boxed them up. This activity continued for the remainder of the week.

Friday morning, Steve brought two new people to Jackie. He explained to her, "Jackie, this is Mr. Ames and Mr. Patterson. They are mathematicians. They are here to help us understand how the brain is wired. It is not a computer, so computer science may not be the answer. They are good at studying new systems and finding the code. I want you to work with them to come up with software to read thoughts."

Jackie grabbed Steve by the arm and pulled him away from the two new arrivals. She had concern in her voice, "Who are these guys? Prima donnas? Why are they here so late in the program? Why me?"

Steve whispered, "You know computers. They figure out systems. Between you, you can find the key. I believe in you." She felt slightly patronized and he saw the expression on her face. He spoke to reassure her, "You have been running this thing since it started. Who else could I send them to? Who else knows what is going on?"

Jackie grinned and answered, "Okay, I guess you are right. What if I don't know what to do with them?"

Steve reassured her, "Then it will be my problem. Just remember that I have to deal with the money people and keep paychecks coming in. You wouldn't want me to get behind on that would you? It is a lot of work begging for paychecks."

Jackie said, "Okay, I'll leave the begging to you. I'd rather just get violent with you instead."

Steve grinned, "I noticed that. By the way, knock it off. I'm really sore from you doing that. Then I have to beg for your paycheck too."

Jackie spoke over her shoulder as she walked over to the new recruits, "You'd better beg for my paycheck." She put her arms over the shoulders of the mathematicians and led them to a console, "Now boys, this is a brain." The mathematicians had bewildered looks on their faces.

On Wednesday of the following week, Jackie pulled Steve aside. There was giddy excitement in her voice, "Did you want to see what a thought looks like?" Steve didn't say a word; he just followed her to the big screen. The screen showed Larry making a pizza. The picture was upside down, distorted, and the movement was fast and jerky.

Steve asked, "Jackie, did you correct for the shape of the eye? How about the speed of the brain?"

Jackie smiled and responded, "No. I just got this part to work. That's next." Steve smiled at her and nodded as he walked back to his office. Four hours later, Jackie burst into Steve's office and shouted, "I think you'd better look at this!" Steve got up and followed her to the big screen. The same memory was playing clearly and with sound.

Steve reached down to Jackie's hand, which was at her side, lifted it up, and shook it. He smiled and laughed, "Yes! That's it! I need lots of copies of that software. I also need one copy of that memory on disk."

An operator stood up from his terminal and tapped Steve on his arm; "This is your memory." He handed Steve a disk in a protective sleeve.

Steve thanked him and as he walked out of the room, he turned around and shouted, "Thank you all!" He returned an hour and a half later and pulled Jackie aside, "I spent half an hour negotiating, or maybe begging, in the president's office. I convinced him with that disk, that proper motivation would get the project finished a lot faster.

I told him that this amount would get the work started, and a promise of more would get it done. You said something about getting the computer to do what a room full of people is doing right now." He reached into his jacket breast pocket and pulled an envelope out. He reached for her hand and placed it in it.

Jackie opened the envelope and her voice sounded like a little girl's voice, "Eight thousand dollars? Eight thousand?" She paused for a moment; "You want a program which traces the neural pathways."

Steve replied firmly, "Yes. Recruit all of the help you want from the terminals. Get any office and office supplies you want. You and your help will get bonuses. Get back to me in one hour with a plan."

Jackie said, "Got it!" and walked around the room, taking operators from the terminals. She led four operators out of the room with a dry erase board and handfuls of markers and paper. She found Steve in twenty minutes; "I have three computer engineers and a mathematician. I took the conference room two doors down. They want this kind of pizza and this kind of beer. I told them I would ask."

Steve spoke and frowned, "Beer? Pizza in the break room maybe, but beer?" He paused, "Clock everyone out and go to a bar. Give me progress reports twice a day. Don't get too messed up. I'll cover the first round." He pulled one hundred dollars out of his wallet and gave it to her. He ordered her, "Get me results!"

Jackie was excited and responded quickly, "Yes Sir! Right away!" She turned and waved as she walked out of the room. Steve heard her shout to the room of her helpers, "Let's go! We are out of here!" In a minute, all five of them ran out of the room and down the hall. Steve heard the laughter and cheering all of the way to the elevator. They bumped the dry erase board into the wall a few times getting it into the elevator.

Steve walked back to his office, sat back in his chair and smiled contentedly. Jim knocked on his office door and raised his voice slightly to get Steve's attention, "Hey Steve, I'm done."

Steve sat up and spoke in a matter-of-fact way, "Okay, I'll see you tomorrow."

Jim cleared his throat, "No. You don't get it. I am finished. I am done cutting the subject. I did the whole thing. I need a paycheck and I'll be on my way."

Steve was surprised, got up to go, and answered, "Okay! I'll go get it. I'll be back in a few minutes. Thanks Jim." Steve walked to the president's office, gave Dave his update, and got Jim's check. Steve returned in twenty minutes and handed Jim his paycheck. Jim smiled, put it inside of the breast pocket of a leather jacket, and then put the jacket on. Steve remarked, "Nice jacket."

Jim answered, "It's my bone bag. If I take a good spill, this should keep me from losing any." Jim saw by the look on Steve's face that he didn't understand, so he explained, "This is my motorcycle riding jacket. If I get in an accident, this is my only protection."

Steve grinned and asked, "You wouldn't happen to have a biker nickname would you?"

Jim replied, "Weeds. When I first started riding, that's where I ended up. I did some wonderful face plants." Jim turned and walked away.

Steve went back to the laboratory to check on progress. The large screen showed fifteen-percent completion. The eyes, reptile brain, and mammal brain showed light gray with faint flashes of light ahead of them. Four operators were still scanning slides into the computer model. The rest of the room was filled with the sound of cooling fans and an occasional hard drive warming up. In the corner, two men were removing the slicing machine, which Jim had been using. The cooling tank was already gone. Without Jackie there, Steve announced the breaks and stayed in the room to handle questions. The project advanced at this rate until Friday.

Friday morning, Jackie and her crew were back. Jackie walked up to Steve and placed two disks into his hand. Jackie could barely contain her excitement, "Go ahead, try it!" Steve walked over to an empty console and placed the first disk into the CD drive. The big screen went blank for a moment. In a few minutes, all of the operators in the room looked up from their stations with puzzled looks on their faces. Every computer in the room was loading the new software. In a few minutes, the big screen showed the brain and completion again. At the bottom of the screen in small print was, "Please insert second disk." Steve changed disks. The screen came to life. The light flashes of new neural connections became so fast that the screen became like a light show. Every operator in the room crowded around the big screen and watched in amazement. Steve went to a phone and called the president down to see the show.

Dave was there in minutes and out of breath. Once he got there, he joined the huddled crowd staring at the light show. While Dave was there, the percent complete advanced from eighteen to nineteen percent. Dave found Steve, "This is what I hoped for! Send your five people to my office. I want a readout of that screen in my office!" Dave turned around and left.

Steve looked over at Jackie and said, "You heard the man." Jackie gathered up her team and went to the president's office. Steve looked around the room at all of the empty computer consoles, and spoke to the group; "We still need ten people to finish scanning the slides into the system and catch things, which the program misses. The rest of you have done a wonderful job and are finished with the project. I will start by being democratic. May I see hands of those who would like to stay?" Nineteen hands went up in the air. The others left. Steve addressed the remaining crew; "I can work ten people for eight hour days, or twenty for four hour days." More people left. Steve counted heads with his finger and spoke to the few who remained "Eleven! That's close enough. We still have work to do. Let's get those slides scanned and keep an eye on the model for loose ends." Everyone returned to work.

In one hour, Jackie and her group returned. She looked around the room and saw the empty stations. Jackie went to Steve's office and asked, "Doctor, what happened to the people?"

Steve glanced up from what he was doing and replied, "We only need a few people now. The others left. You and your group are still on the project." She nodded and walked back to the laboratory. Steve got up from his desk and went to the lab to see the progress. The room was surprisingly busy. Fifteen people were scanning slides into the computer. One person was searching the computer model for gaps in coverage. Only four stations could scan the slides, but everyone else was handling the slides, placing them in order, and boxing them for storage. All of the boxes of slides could have fit in a small outdoor ten by ten shed. An object, which originally weighed three pounds, now weighed over one ton. The used slides accumulated at the other end of the room, where the slide slicing machine and tank had been. The display on the big screen looked like an almost complete human brain. The portion, which was already traced, was now twenty six percent.

CHAPTER 5

WHAT THOUGHTS LOOK LIKE

I t was two thirty in the afternoon. The phone in Steve's office rang and he was down the hall, so he heard it. Steve ran to answer the phone; "Hello Doctor Warren.

This is Detective Edward Morris. I haven't heard from you."

Steve answered, "Yes detective. I remember. You want to know about the project."

Ed replied, "Yes. Do you have anything to report?"

Steve was excited, "I think you should come down here and see for yourself. I don't know if we can collect anything yet, but I know it will be worth your time. Do you know how to get here?"

Ed said, "Just fax a facility map. I know where Fitsimmons is. I'll be there in half an hour."

Steve answered quickly, "Done! Bye!" He pulled the detective's card out of his wallet, rummaged through his desk drawer for a map, marked it with a blue marker, and fed it into his fax machine. He waited until he received the confirmation, then he rushed to the lab and addressed the group: "Hey everyone! We have a VIP coming in half an hour. Can we collect memories for him? Recent memories?"

Mike was at the terminal and searching for gaps. He spoke up: "Some of the new ones, yes. The old ones aren't traced yet. I'll see what I can find. Did you need copies?"

Steve turned and walked out of the room and replied over his shoulder, "Yes, make it three."

In thirty-five minutes, Steve met Ed in the parking lot and led him to the laboratory. Steve introduced Ed to the room; "This is detective Edward Morris. He is here to get a statement from Mr. Curtiss. Do we have some video for him?"

Mike looked up from his terminal and answered, "Here's a good one boss." The big screen showed thirty one percent complete and still flashed wildly. The normal display of the brain moved to the right and became smaller. The upper left corner of the screen began to display a scene of Steve's house in the dark. The room was quiet except for the sound of the memory playing. Everyone watched Larry break glass at the back porch window and muffle the sound with a piece of cloth. They all watched him move around the kitchen. Larry looked around the kitchen and took a large knife from a drawer. His thoughts played out loud; "This will make it look like some punk tried to rob the Doctor and killed him in his sleep. I'll do them both." Larry slid the knife into his back pocket and continued to walk through the downstairs part of the house in the dark. Larry approached the base of the stairs and saw a dark shape. The shape looked like a man. His head looked down to find his pistol and there was a bright flash. Larry's head movements became clumsy and the screen began to fade out of focus. Another bright flash of light followed and the picture became dark. There was a low pitched sound, which became louder. It was a mournful and moaning sound. It was painful, not to the ears, but to the soul. Dark, slowly moving shapes became visible on the screen. Outstretched arms and hands swayed back and forth. Faces, barely visible, with mouths wide open were everywhere. Arms and hands everywhere around Larry pulled him lower and lower. The movement on the screen was like a floating feeling. Random shapes moved around as if they were swimming. Pain and anger showed on faces all around Larry. Hands in violent poses moved in a slashing motion toward Larry and indications of pain flashed on the screen. Red and purple streaks moved across the screen until everything was black on the corner of the screen, which showed the memory. Everything was quiet throughout the room.

Steve and Ed broke the silence, "Give me two copies!"

CD writers across the room stopped and CD carriages opened up everywhere. Fifteen hands held CD's up in the air. Mike collected four and handed them to Steve and Ed.

Steve spoke up; "I guess Larry was a bad boy. That doesn't look like heaven to me."

Ed remarked; "I'm always glad to see bad guys end up there. This will clear your name doctor. I just hope it is admissible in court."

Steve chimed in; "It will be. The brain is like any other video recorder. Even better than that, the victim has no rights. The Bill of Rights doesn't apply to the dead."

Ed smiled; "While we are questioning a dead man, let's see what else we can get from him." He slid the two disks into protective covers and put them into the breast pocket of his suit coat. Ed carried an air of professionalism around with him. He was a tall man in his forties with a very short hair cut. He was alert and serious at all times. He seemed to be the kind of person who could scuba dive in a muddy lake and come out clean and without any wrinkles in his suit.

Steve spoke to Mike; "Let's go a bit further back in the short term memory."

Mike answered quickly and raised his hand like a piano maestro playing an exciting piece of music; "This is my encore performance!" His hand went down on the mouse and the big screen became bright with the new scene. Steve's wife was on the screen.

Steve and Ed both shouted; "Back that up!"

The screen backed up almost like a VCR and the scene returned to when Rose was lying down and speaking softly; "I already paid the network for the job. Steve needs to die tonight. I found out he's been hiding a lot of money in a lot of accounts. If I am the widow, I can have it all to myself."

Larry's voice spoke, but Rose still filled the screen, "Don't push! I am a professional and I know what I am doing. Whatever happens, you don't know anything. Don't make me come back and deliver a freebie."

Rose backed away and held the blankets to herself as if they would protect her; "I hired you and you would kill me?"

Larry answered in a way, which did not sound like Larry. His voice became hard and cruel; "Lady, in this business, we don't leave witnesses. We don't answer questions. We don't sleep with customers either. I already broke one rule. Don't make me tie up loose ends. I don't dislike Steve. Don't make me dislike you. You didn't hire me anyway, the network did. Don't tell me how to do my job!"

Rose backed away with a frightened look on her face.

The screen went directly to Larry making pizza.

Mike spoke up; "You don't have to say it! I made six copies!" He got up from his seat, and went from terminal to terminal, collecting disks and placing them in sleeves. Jackie took them from his hand and gave them to Steve and Ed.

Ed spoke to Steve; "You realize that we will be placing your wife into custody now. Don't bother going home for awhile." He tucked his three copies into his breast pocket.

Steve was furious. His fists and jaw were clenched. It took effort to speak, "You'd better before I kill her!"

Ed spoke seriously, "Don't go home. I will talk to you again. Where will you be tonight?"

Steve was distracted. His thoughts were racing; "I don't know. I have to leave for a little while. Give me a few hours. I will call you." He looked at his watch and thought for a moment. He then said; "I will call you at six. I have to go. I really have to go."

Ed warned Steve, "Don't go home. Don't contact your wife."

Steve answered, "Home? No way! Contact her? I don't even want to! I have to go now!" He walked into his office, pulled the discs from his pockets, and left them on his desk. He walked out and left his office unlocked.

Jackie turned to Ed and asked, "What do you think he's going to do?"

Ed answered, "Hard to say. Some people go to the gym and work out their stress. Some people go to bars and start fights. Some people go somewhere to be alone and calm down. Whatever he usually does when he is mad is what he will do."

Jackie asked, "Do you think he will do something really bad?"

Ed replied, "He doesn't have a criminal record. He is over forty years old. If he had behavior problems, he would have gotten in trouble before. I don't think so."

In just a few moments, Steve opened his car and got in. He started his car and sat there just staring at the gages for a moment. He shouted, "Rose! The counseling worked! You got worse! I just wanted a divorce. I guess you don't stop there do you?" He put the gear selector in drive and left the parking lot. Occasionally, he pushed the accelerator just to feel the engine. He didn't leave a quarter mile of dust, smoke, and marks like he wanted to. As he drove out of the parking lot, he noticed a black Dodge Ram pickup. It made the same turns he made.

Steve began to watch the truck. It was slightly raised and had large tires. The eight-foot bed had flared wheel wells, so he knew it was a heavy-duty model. The truck kept its distance. As Steve turned left, out of the Fitsimmons complex onto Colfax Avenue, he watched to see what the truck did. The truck turned in the same direction. Steve felt rage and fear. His heart pounded in his ears and stomach. The Dodge stayed several cars behind, but always within sight. Steve spoke to the driver of the truck, even though he knew that the driver couldn't hear him; "Let's see if you are just going in the same direction or following me."

Steve led the truck away from where he wanted to go.

He drove very slowly and the truck passed him. He then turned into a neighborhood to lose the truck. After a few turns, he left the neighborhood, got back on Colfax again and the truck reappeared. Steve spoke again; "If you want to play, then we will go to my playground." He led this tail to his special road.

Steve planned his move. He kept a speed of twenty miles per hour and watched what the truck did. At this speed, anyone who wasn't interested in Steve would have passed him. Twenty miles per hour was also slow enough to make a sudden get away, if he had to. The Dodge made contact with his rear bumper and pushed the light car forward hard. Both cars reached twenty-five miles per hour. The bumpers were grinding and thumping each other violently. Steve was filled with a violent rage and shouted, "My turn!" He goosed the accelerator hard. The vehicles separated. He turned his front wheels left hard, and the rear of his car made a large and quick circle. As Steve's rear end made its arc, the truck bumper caught up again and the car bounced off. The extra momentum spun the car around faster. Steve kept his wheel turned left and he just managed to hit the left quarter panel of the truck. The blow was hard enough to point the truck toward the opposite ditch. The quarter panel of the truck was the sweet spot. It was the only part of the vehicle that Steve's light vehicle could move. The driver of the truck tried to turn away from the ditch and stop the truck, but he was in four wheel drive, and none of the tires had traction on the loose gravel. The front tires of the truck fell over the sharp shoulder and the truck body hit the ground hard. It was high centered and couldn't move anymore.

Hitting the quarter panel of the truck slowed the Skylark down enough that Steve could regain control of his vehicle. He missed the ditch and stopped his vehicle with the truck in his rear view mirror. Steve watched the truck window roll down. A hand with a pistol emerged. Steve shoved the gear selector into "R" and aimed for the spot between the truck bed and cab. The hand retreated back inside the truck window.

Steve stopped his car, opened the door of his car and ran toward the tailgate of the truck. He jumped into the truck bed, looked around to see if he could use anything there, and found a chain. He took the end of the chain and measured an arm length. He held the chain one arm-length from the end and stood ready. The other man heard him moving around and taking the chain. The driver of the Dodge truck quickly unlatched and kicked his passenger door open. He then opened the driver door. The assassin vaulted from the truck on the passenger side, rolled on the ground, recovered with his pistol in front of him, and fired one shot at Steve. The shot missed, but it was close. Steve jumped to the back of the truck and swung the chain at the man. Steve kept his hand around the chain, but released his grip, so it could feed toward his target. The chain hit the other man in the face, and he grabbed it to pull it away from Steve. Steve ran off of the end of the truck, holding on to the chain, which was now anchored to the bad guy. The assassin suddenly realized that Steve was using him as leverage, and released the chain. When the assassin released the chain, Steve used the momentum of his body and the release of the chain to swing the chain in a low arc across the other man's chest. The other man ducked, but instead of the chain missing him, it caught him around the neck. The weight of the chain knocked him down. Steve followed the chain and delivered a strong kick to the left side of the man's head.

The pickup driver didn't move. He was out cold. Steve used the chain and pulled him toward the truck. He pulled his head against the truck body and wedged the chain between the truck and ground, so that if he did regain consciousness, he couldn't escape or attack him again. Steve searched the man for identification and weapons. Steve found a cell phone and wallet. He picked the pistol up off of the ground. He tucked these things into his pockets. Steve jumped up into the bed of the pickup and found a fresh new roll of duct tape. He spoke out loud; "Who were you going to use this on pal?"

Steve jumped out of the bed of the truck and opened the roll of tape with his teeth. He restrained the hands of the assassin with the tape and then released him from the chain. He pulled him to the rear of the truck and wrapped the tape around the bumper and his hands. He worked the upper body of the bad guy up to the bumper and lashed him there. Then he began taping the feet to the bumper. The assassin woke up and resisted. Steve walked around the side of the truck and took the chain again. He used the chain to pull the other foot to the bumper. Then he began to complete gift wrapping the bad guy for the police.

Once he was done, Steve pulled the wallet out of his pocket; "Mister Turner. Aw heck, why don't we just say Smith? Oh, this is rich, Elm Street! That's good! Are you the nightmare? What do we have here? You work in a grocery store too? Let's just drop the act pal! I know why you are here." Steve moved closer and whispered; "I'm going to call a police detective on your cell phone. He will come and take you away. This is personal! If I see you again, I will give you a frontal lobotomy. That will end your fun forever. You must understand; I am having a really bad day and I want to take a road flare out of my trunk and incinerate you. Can you give me a reason not to overreact?" The other man didn't know what to say. Steve said, "You aren't helping at all!"

Steve reached into his wallet and found the detective's card. He dialed the detective's number; "Hey Ed, I will give you directions in a minute. You'll never guess what happened to me." Steve filled the detective in on the details and gave him directions. The police arrived in an hour. Ed took the man into custody and sent him away with a squad car. Ed turned to Steve and spoke inches away from his face; "Mister Warren, you have acted without authority and ruined a crime scene."

Steve was getting angry; "That's Doctor."

Ed growled, "I'm looking at a sixteen year old kid with gray hairs who wanted to play demolition derby on a back road. I don't know any doctors who drive reconditioned junkyard specials and play chicken with pickup trucks. Are you reliving your youth, or didn't you get around to having one in the first place? Something tells me that this is not acceptable behavior for a Doctor of Neurology. Playing motor head in your free time is your business, but when I have to spend taxpayer money to come out here and clean up your mess, I have the right to say something. I don't know how long we can keep this Mister Whoever. You unlawfully restrained him, unlawfully searched him, and attacked him with a towing chain."

Steve spoke up; "It was self defense. He shot at me!" Ed replied, "With the pistol which has your prints on it."

Steve was angry, but held back; "I was keeping it away from him."

Ed snapped back, "You don't know how lucky you are that I believe your story. I have enough to keep you in a cage for over a month. Your own statement is incriminating. The damage to the vehicles proves your story. The man we hauled away could sue you for damages. He could charge you with assault and simple assault."

Steve was puzzled; "What is simple assault?"

Ed explained, "Simple assault has two parts. The first part is making a threat. The second part happens when the victim believes it. You told him you'd give him a frontal lobotomy. You are a neurosurgeon. You could! Watch your mouth. You are getting in deeper by the minute."

Steve backed off, "Yes sir."

Ed calmed down; "I know that you have had threats on your life. I really do believe that this man chased you with murderous intent. I know you are scared and now you have a divorce to think about too. At the same time, something really big is happening in your career. Have I left anything out?"

Steve answered, "No Ed. That sums it up pretty well."

Ed pulled a couple of cards out of his breast pocket, "Go get counseling, get a lawyer, and get to work. If you see something suspicious, get home or to your office and call me. My people will handle it. I'm willing to bet that you already know a person you could start a better relationship with, so start dating again. Concentrate on that. Sometimes, work is therapy. Dig in and get busy. I'll make you a deal. I won't perform any surgery without a license if you don't attempt to arrest bad guys without proper authority."

Steve was worried; "You said this guy might go free. Can't I defend myself?"

Ed answered, "We will watch him at a distance once free. He won't go anywhere for twenty four hours from the time he is booked. I have some reckless driving charges to handle with him, which may keep him longer. This is Friday, so he is stuck until at least Monday. It depends on his lawyer. I will call you when he is about to be set free. Fair?"

Steve said, "Yes. Thank you. What if I leave the State to flee?"

Ed smiled; "Then you are someone else's problem. Don't go home yet. Two units are on the way now."

Steve smiled; "Then I'm going to a bar. I'll be home very late."

Ed smiled as he turned and walked to his car.

Steve returned to the laboratory by five-fifteen. Everyone was still there. Jackie walked up to him and looked concerned; "Is everything okay?"

Steve smiled; "It's more than okay. I got a good workout. I feel much better." Steve looked up at the big screen. The percentage complete was thirty-four percent. Steve addressed the entire room, "Team! You have done a wonderful job. In three weeks, we are a third of the way finished. I want to go to a bar. Would anyone care to join me?"

One of the operators stood up and addressed Steve; "Sir, I have to go home. My family is expecting me."

Steve smiled, "Then don't let them down. Have a good weekend." Steve turned and walked toward the door. The room emptied quickly. The display on the large screen advanced to thirty-five percent. In small letters at the bottom of the screen, a message appeared; "Where am I? I can't feel anything. Is this Hell?"

CHAPTER 6

LIFE

The team met in the parking lot. Jackie spoke to Steve; "Which bar are we going to?"

Steve shrugged his shoulders; "I haven't gone to any in years. Why don't you pick one and we can go in a convoy. Any better ideas?"

Jackie looked around at everyone and said, "Everyone okay with the Scottsman?"

Everyone looked at each other and shrugged. Mike asked, "Can we eat first?"

Jackie smiled; "We can eat there. The smothered burritos are good. If there aren't any better ideas, follow me."

The group separated and went to their cars. Steve spoke up, "Jackie! I don't want to drive. Can I ride with you?"

She smiled and pointed at her passenger door; "Hop in!" Steve went around her vehicle to the passenger side. Before he could get to the door handle to pull it, he heard the lock tumbler click. He pulled the door open and got in;

"I'm not used to that."

Jackie gave him a confused look; "Used to?"

Steve answered right away, "The new gadgets. Electric door locks. You know."

Jackie was really puzzled, "Why not? You are a big time surgeon with big bucks. What do you drive?" Steve pointed to the car next to them. She shrieked, "Wow! Why?"

Steve quietly responded, "Because I built it and I like the power. I might get a new car too just so I don't have to hear that anymore."

Jackie apologized; "I'm sorry. It is macho. I just would have expected a medical student to be driving it."

Steve was sarcastic; "Gee thanks! That's even worse.

Knock it off. That isn't why I wanted to ride with you." Jackie asked, "Why did you?"

Steve stammered a little, "I don't drink and drive. I also wanted to ask you if you wanted to spend more time with me. I'm working on my first divorce and I like you."

Jackie answered, "When did this happen? Are you asking me out?"

Steve was embarrassed; "Yes but only if I don't get hammered with sexual harassment. I have enough problems already."

Jackie blushed a little, "Okay. I won't hammer you, but if you get out of hand, I'll pound you into the dirt."

Steve joked, "Should I run now?"

Jackie smiled, "I'll tell you to run first. Until then, you have a chance."

Steve relaxed, "That's good. I feel better. You wouldn't be mad at me if I told you I like the way you do things? What if I told you I like the way you look? I think you have a wonderful smile too."

Jackie responded, "Thank you. I'm not mad yet. Tell me.

Why did your wife want to kill you?"

Steve looked away, "We hate each other. We have practiced that for twenty-four years. We know exactly how to make each other mad. We each want exactly what the other isn't. We are beautifully dysfunctional."

Jackie asked, "And what are you?"

Steve looked back, "Busy at work and getting older." The rest of the conversation was friendly. Steve and Jackie shared a table at the bar. At eleven-o-clock Steve and the rest of the crew were curious about the progress of the project and decided to return to check again before going home.

Steve was the first one in the laboratory. He saw the display on the screen and asked everyone in the group; "Is someone here playing pranks?"

Nobody said a word. Michael Owens went to a terminal and wrote, "Who are you?"

The answer was; "Larry Curtiss. Who are you?" Mike wrote, "Doctor Michael Owens."

Larry asked; "Am I sick?"

Mike said under his breath; "No, you idiot! You are dead!" He typed, "No, but why are you calling yourself the name of a dead person? Is this a joke?"

Larry answered on the speakers of the computers, "This isn't a joke. I am dead? Then that is why I see my life flashing before my eyes. What has happened to me? This feels strange. Please explain."

Mike typed, "You died and we began to build a computer model of your brain in this machine. I don't understand why this is happening. You should not be able to interact with us. I don't understand why I am able to talk to a dead man."

Larry spoke loudly; "You don't understand? I am seeing through black and white video cameras. Old memories are going through my head. I guess I don't have one anymore. I am reliving my whole life. I woke up into this? Where is my body? Is what you did legal? I am guessing that I am a laboratory experiment now. I see Doctor Stephen Samuel Warren. I want to talk to him!"

Steve found a terminal and typed; "You wouldn't happen to have a wild garlic pizza in there would you?"

Larry answered on the terminal that Steve was typing on; "Yes I do! Why don't you come in here and get it?" A memory of Larry making a pizza flashed on the screen. Larry wrote, "Bring the beer and we'll hang out for a while. I looked ahead at the last memories. I couldn't see you very well, but I am guessing that you killed me. Maybe I should be mad at you."

Steve wrote, "You have that choice. I am already mad at you."

Larry paused for a moment and then spoke on the speakers, "I can see your point, but the person you need to be mad at isn't in this room. I'm fuzzy on the time frame, but I'm guessing that it's been almost three weeks. Why are you still alive? There should have been another hit."

Steve wrote and it appeared on the big screen; "That was this afternoon. A black Dodge Ram pickup tried to run me off of the road."

Larry asked, "Did he have a camouflage cell phone cover, duct tape, and a towing chain?"

Steve answered; "Yes! You know him."

Larry laughed over the speakers; "That is the rain man.

They sent him?"

Steve replied, "I'm not laughing."

Larry spoke loudly; "You don't send someone who has trouble with little old ladies after an ace. The guy is a knucklehead. He has a rotten attitude, got caught twice already, mouthed off at the network, and is very sloppy."

Steve wrote, "What is an ace?"

Larry answered, "You are. You whacked a hitter. Next time they will send a team. No way you get to live. They won't take any chances with you. If they sent the rain man, then they wanted you to kill him for free before they come after you again."

Steve asked, "How long will that be?" Larry answered, "About a week."

Steve asked, "Can you give me a reason not to turn this computer off and kill you again?"

Larry spoke, "I understand Steve, but I am not the reason your life is in danger."

Steve pounded the keys hard, "But, you are a murderer!" Larry answered calmly, "Don't think of it that way. The customer is the murderer. I am the weapon. Whether I complete the contract or someone else does, the customer made it happen. There are fifty people who could come after you. I am the least of your worries, and I could help."

Steve pounded the keyboard so hard that one of the keys bounced up from the keyboard and fell to the floor; "Why would you help someone you tried to kill and who killed you?"

Larry said, "I think you saw my after death experience." Steve replied, "I laughed through the whole thing!"

Larry spoke, "I felt the whole thing. It is worse when you are there, and I don't want to go back. If I make a few things right, then maybe I won't go back. I can start with you. Now that we have this new and close friendship going, you can make the rest of my new life better."

Steve shouted as he wrote, "Friendship? I hate your guts!"

Larry spoke with reassurance, "Buddy! You are going to like me. I am going to save your life. I have talent and looks; well I did have looks. I do have charm though. You need me and I need you."

Steve calmed down; "Why do I need you, and what do you want from me?"

Larry spoke with a chuckle in his voice, "Because I am the best hacker in the world. I am not at a keyboard. I am in the machine. Every second of your life is several hours of mine. I swim around in here like a shark and programs look like small red fish. If I touch one of those fish, I know everything about it. I have never been smarter in my life. If I had known all of this while I was in school, I wouldn't have chosen my last profession. I can save your life because I know how. Once that is done, I want a nice big computer to live in, online access, a few electronic toys, and a landlord. Don't worry, I can pay my way."

Steve marveled, "How can you pay your way?"

Larry answered, "Do you know my real name yet? If not, then nobody else does either. The bank doesn't know I am dead, and I can send an electronic check for your last operation on me. Would one hundred thousand dollars cover my first six months in your home? If anything happens to you now, I am a goner anyway. Are we friends yet?"

Steve wrote, "Yes"

CHAPTER 7

FRIENDS

Harry spoke loudly again; "Good! Now that we're friends, get to work. Go get an ear from the morgue, unless you want to send your own. Get a can of fresh coffee from a supermarket. Get an old rag and make it bloody from the ear. Get some zip lock bags, a box to mail the coffee can in, a set of your own fingerprints, any current death certificate, and pictures of a mangled body which could be mistaken for you. Place your driver license on the scanner now and once you have it, the death certificate. Hurry up! You only have until noon tomorrow to get it in the mail. By the way, I have seen your handwriting. Get your girlfriend to address it for you."

Steve was confused; "Girlfriend?"

Larry chuckled, "My vision isn't too good anymore, but I have seen the way you look at the woman over there, and the way she looks back. Get to work!"

Steve answered, "I'm going." He turned to Jackie; "Would you ride with me please?" She just nodded and followed. They got to the parking lot and Steve led her toward his car. The expression on her face showed pain as she approached the beast. She looked at Steve with pleading in her eyes and shook her head, "No." Steve looked at her and said, "Come on. Knock it off. Will you just get in?" Steve opened her door and walked around to his side. He opened his door, got in, flipped the switches, started the engine, and turned on the lights. The interior of the vehicle looked like the space shuttle. Steve placed his hand on the metal part of the dash and then put the gear selector in drive. He barely touched the gas pedal and the vehicle lurched forward with great force. He wasn't going very fast, but Jackie was holding onto the seat belt with a death grip.

Jackie asked, "Why did you touch the dash before you put it in gear?"

Steve replied, "I can't always hear it run. I can feel it though." He glanced over to Jackie and looked back at the road, "Jackie, Why was I talking to a dead man back there? How was it possible?"

Jackie spoke, "I think it was the writing program. The big program, which connects the dots. It turns the connections on to see if they work. The testing part of the program thinks like Larry. He is inside the thinking part, the RAM and processor, of every computer, which is running in the room. When you turn a computer on, the RAM, or Random Access Memory, fills up with programs from the hard drive, which is the long-term memory of the computer. The RAM serves the processor, which does the thinking and it is only awake while the computer is running. The writing program I gave you is so big that there is very little room left in any one computer for Larry to think with. That means that he is spread over several computers. Larry didn't speak until now, because he didn't have enough of his brain mapped out in the computer to do that. If there were only one computer on, then Larry couldn't even think anymore. Larry can't touch the map we are making. He isn't online either. Do you trust him?"

Steve answered, "I trust him up to a point. I see where he is going with this set of instructions. He wants me to mail an ear from the morgue to the network; whatever that is. I think he will e-mail a phony death certificate too. That will fool the bad guys for a little while. I'm still waiting to find out why I need him as a hacker. I have a feeling that I won't like the next thing which happens."

Jackie looked worried; "I guess you won't. Would you mind slowing down a little? I didn't sign up for this job to race at Indy."

Steve was confused, "Jackie, I would, but if I do, I'll get pulled over. I drove slowly to calm you down. I'm only going thirty. The speed limit is thirty-five. You just aren't used to this car. Look." He pointed at his speedometer; "I know this feels different than your car, but then, your engine turns twice as fast as mine does, your car doesn't have a rigid frame, your vehicle has more soundproofing, and you can't spin your back wheels. This is as different from your car as a school bus."

Jackie kept her firm grip on the seat belt; "If you say so." They pulled into a grocery store and collected the coffee, dish towels, zip lock bags, and a box to mail everything in. On the way back, Jackie told Steve, "He can go online, but he can't leave the lab without a different version of the writing program. If he could leave the computer at the lab, then he would either be the writing program, or the map. Both can't go anywhere together. He is stuck in that computer until we let him out."

Steve said, "I hope you are right." When they got back, Steve left Jackie at the laboratory with the supplies and went down to the morgue. He returned with the ear, fingerprints, ugly photographs, and a death certificate.

Larry told them how to package and address everything. He then told Steve to get him online. Jackie turned the terminal on which had online access. Before she could do anything else, Larry signed in under his own account, signed in with his bank and sent an electronic check to Steve, sent three e-mails to friends, and sent an e-mail to a strange address. She recognized the message, but didn't have time to read anything. She saw the screen scroll down through the driver license, death certificate, and photographs. Larry then erased his sent file from the mail server. Everything was sent and erased before she could save it or print it. The computer then signed out and shut down.

Larry announced, "Doctor, as soon as my friend redirects the e-mail, it will look like the rain man sent it. You are temporarily dead; just long enough for us to shut the network down. In three hours, you can send the package. I just left a message for another friend. That friend is holding something special for me. I told her that you would come by to pick it up. We need this special piece of gear to do this job. I am printing up directions on that printer. Go there after you drop off the package."

Steve asked, "What are we picking up?"

Larry answered, "Something I once used on a special job. My customer wanted to watch the mark die in real time a thousand miles away. It is a cell phone, laptop computer, and video camera, all joined together and powered by a twelve-volt car battery and a power converter. It can charge from a car cigarette lighter outlet. It looks like a piece of junk, but it was hard to get everything to work together. The customer was able to point the camera, zoom in, and out, from his computer."

Steve, Mike, and Jackie went together to send the package and get the gadget. When they got there to get the gadget, both Steve and Mike had to carry it down two flights of stars from an apartment. It was in a wooden crate with wires sticking out. A tripod with a camera was bolted to it. The whole mess stood less than three feet high. A card with instructions was taped to the tripod. It didn't fit in the trunk of the beast, so Steve fastened it into the back seat with a seat belt.

They were back at the laboratory before nine fifteen. Larry greeted them as soon as they walked in the door, "Took you long enough. Did you mail the box and get the special package?"

Steve answered him out loud, "Yes Mom! Anything else?" He walked over to a terminal and typed a reply, "Yes!"

Larry echoed over the speakers in the room, "Let's party!

We are going to Medicine Shoe, Nebraska."

Steve looked at Mike and Jackie with a confused look on his face. They looked back with a hint of panic in their eyes. Steve slowly typed his question on the computer keyboard, "Why?"

Larry responded thunderously, with a hint of excitement in his mechanical voice, "Because that's where the bad guys are! You don't think they are gone do you? We just got them to look the other way while I do my dirty deed."

Steve was annoyed because he didn't like surprises. He typed, "I'm starting to dislike you again! You just love getting me in trouble don't you? What dirty deed?"

Larry answered over the speakers, "Your part will be to get the bad guy's computer signed online and your file pulled up. I'll take care of the rest. Don't worry. I'll do the hard part. Haven't I been giving you good advice lately?"

Steve paused for a moment and then answered, "Yes." He mumbled, "What am I getting myself into?" He typed, "Why can't you do the hacking from here?"

Larry answered, "Because the jerk is too smart to keep the files on his own computer. He hides the stuff somewhere else. He has to call up the web site for me to find it. If this guy weren't so smart, about ten of us hit men would have waxed him already for some of his dirty tricks."

Steve turned to Mike and Jackie. He grinned sheepishly as he asked them, "Anyone up for a road trip?" Mike and Jackie looked at each other and then back to Steve.

Larry spoke over the speakers, "We can leave as soon as you get me online." Jackie walked over to the online terminal and turned it on.

Steve shrugged his shoulders as he watched Jackie. He spoke to Mike and Jackie, "I guess it's time to leave now."

Larry reminded them before they left, "Steve. Plug the gadget into your car. The instructions are taped to it."

When they got to the parking lot, Steve led them to his car. Mike spoke up, "We aren't going in this are we?" He looked at Jackie for support. She was quiet. Mike spoke again, "We are just getting the thing out of your car and putting it in mine aren't we? My car would be more comfortable for the ride. It has all of the comforts of home. Aw, come on Steve."

Steve looked at Mike and smiled as he explained, "If we run into trouble, how would your Lexus do against a pickup truck?"

Mike looked worried and asked, "Trouble? What are you getting us into?"

Steve walked over to Mike and faced him. He spoke in a low tone of voice, "Are you afraid? Still want to use your car? Would you rather I had an Escort? I've already dealt with two of these guys. Have you?"

Mike backed away and reached for the back door handle of Steve's car. It was locked. Mike frowned as he spoke to Steve, "You win! Now why don't you unlock this thing so we can get it over with. I feel stupid riding in this ridiculous heap. Grow up Steve. News flash! You aren't sixteen anymore."

Steve grinned and told Mike, "Okay, you're right. Let's take your Lexus! I'll drive."

Mike looked at the body damage on the Beast and replied, "Never mind. Now will you unlock this thing?"

Steve turned and walked to the driver side door. He unlocked his door and the other doors. He opened the windows and pulled the instructions from the side of the gadget. All the note said was, "Plug into the cigarette lighter and push the power on switch." Steve pulled the connector out of the box and plugged it into the cigarette lighter. He pressed the power button on the laptop computer.

For a brief moment, the only sound was a low-pitched hum as the hard drive warmed up. The hum stopped and the gadget began to move. The camera elevated and turned toward Steve. Larry spoke over the speakers on the laptop, "I'm back." The camera looked around the whole car and Larry commented, "I was joking about taking your car. Another car would be fine. Don't you have one with a nice stereo?"

Steve reached for the keyboard, but spoke out loud first, "Don't push it! I know where the power switch is knucklehead!"

The camera looked back at Steve and Larry answered him before he could type, "I heard that! I am precious cargo and the navigator on this trip. I would prefer a car with a softer suspension, so the camera is more stable, but this will have to do. The camera has a microphone, so I can hear you. I'm not the one who married your wife. Who is the knucklehead now?"

Steve growled, "You could be replaced by a Walkman." Larry answered with a chuckle in his voice, "Can a Walkman do this?" A small round panel under the lens moved forward and back. It looked like a tongue sticking out.

Steve laughed and replied, "I can't top that. Okay, let's go Cyclops." Everyone strapped in and the car began to move.

Larry spoke up, "Miss, I don't know your name. Would you take the laptop into the front seat? Don't worry, the cords are long enough. I'm putting a map on the screen."

Jackie reached back and pulled the computer forward.

She answered Larry, "My name is Jackie."

The sun was bright and the whole world seemed to sparkle. A few small clouds overhead made bright spots in the sky. Steve put his sunglasses on and so did everyone else. When they reached the interstate and left the Denver area, the air rushing through the car lifted everyone's spirits. There was a feeling of adventure and excitement. Steve turned the stereo on and put in a cassette tape. Pink Floyd filled the car with, "The Wall." Jackie joined in and sang, "We don't need no thought control." At first, Mike grinned mockingly, but then began to sing too. Larry added his voice to the horrible racket.

Hours and miles passed. Late that night, the car pulled into the small town of Medicine Shoe. Steve pulled up to a diner and everyone went in to get something to eat. Mike reached for his wallet after he ordered his food and Steve stopped him. Steve told him, "This is my party, I'll spring for the food. Maybe you can find a map of this place in a phone book."

The waitress interrupted, "Here's one." She handed him a napkin with a map printed on it. She cheerfully explained, "This town is small enough to put on a napkin and you wouldn't believe how much business these things bring in here."

Steve smiled and asked, "That's very helpful of you. Would you show me where the post office is and maybe a motel?"

The waitress smiled and pointed at an intersection on the napkin. She cheerfully replied, "That's the post office, but the nearest motel is in Lantern Peak, fifteen miles away. The nearest K-Mart is there too. Did you come all of the way here just to use our little post office?"

Mike chimed in, "Geologists. We are digging nearby and want to mail samples back to the lab."

The waitress had a puzzled look on her face and asked, "Wouldn't you like to wait for Monday when they are open?"

Mike said, "We are. We will. We just came from the dig to check out the town and see what is available. We thought it might be nice to have some of your world famous bacon biscuits." Mike pointed up at a sign on the wall.

The waitress giggled, "World famous. That silly sign? The owner made that dumb thing. It's a local joke here." She chuckled as she walked away.

Steve whispered to Mike, "What are we going to do here until Monday? Go to Lantern Peak?"

Mike cheerfully responded, "Yes and go see the world famous gangster museum." Mike held up the napkin and pointed to a picture in the corner.

Steve reached for the napkin and looked closely. He spoke with surprise, "The what? Who? Could it have something to do with my little problem?" Steve laughed and looked at Mike as he joked, "I like how everything here is world famous."

Mike grinned, but tried to sound serious, "People all over the world come here for the world famous biscuits. Howard Hughes had his at this very table." Mike reached over to the napkin, pointed at another corner, and joked, "We must see the world famous beer mug museum and largest working phonograph record in the world. The record player weighs 467 pounds."

Jackie chimed in, "Oh yes, can we?"

Steve grinned and said, "I'll show you a beer mug collection."

Jackie looked shocked and scolded him, "Hey knock it off!"

Steve pointed to a corner of the diner and said, "No look!" Mike and Jackie both looked and laughed.

Mike laughed as he spoke, "I don't think that one is world famous."

Jackie chuckled and replied, "Are you sure?"

They ate their food and left. That night, they drove to Lantern Peak to stay at a motel. Steve handed Mike some money and told him, "You need to put the rooms in your name, because I have to remain anonymous here."

Mike went into the office to get the room keys and returned shortly. He handed the keys to Steve and Jackie and said, "Steve, you are my Dad, Oliver. Jackie, you are my sister, Brenda."

Larry interrupted, "Where are my keys?"

Mike looked at the camera lens and asked, "Where is your 115 volt power cord?"

Larry answered, "What? Oh no! But." Mike pulled the plug out of the cigarette lighter and the gadget became silent. Late, the next morning, all three met outside of their rooms. Steve started the car and plugged the gadget back in. The camera elevated and swung around from side to side. It turned and faced Mike. Larry yelled, "You rude, no good, jerk! You don't just pull the plug. You knucklehead! You have to shut it down correctly. Don't you know anything? You could have damaged the computer!"

Mike sneered as he spoke, "Only if we're lucky."

Larry answered, "Messing up this computer would be the same as making this trip for nothing. Can anyone here hack the network computer?"

Steve spoke apologetically, "No, I'm sorry, we can't."

Larry ordered Steve, "Go to a store and buy a video camera. Make it eight-millimeter format. Keep the room keys because you may need to charge the camera battery. Now move it!"

Steve was annoyed because he didn't like being ordered around. His attitude changed and his voice was defiant, "Why this time?"

Larry answered somewhat patronizingly, "Because I can't go into the museum and I need to see what it looks like inside. By the way, do you have a weapon?"

Steve was surprised as he answered Larry, "No, the police have mine. Do I need one?"

Larry responded quickly, "You might. Reach under the power box of the gizmo. I have one taped in place. Now hurry up and get the camera."

Steve gave up and told Larry, "Okay, I'll do it now." Steve motioned everyone to get in the car. They went to get the camera and followed Larry's instructions. A few hours later, they arrived at the museum. Steve stopped everyone from going into the museum. He told Mike and Jackie, "Just a minute." He reached into the wooden crate and under the power converter to get the pistol out. Steve tucked it into his sock.

Mike watched him do this and asked, "Did you know that was there? Is that a good place to put it?"

Steve answered, "I don't know. I don't usually carry one of these things. Do you have any better ideas?" Mike shrugged his shoulders and walked toward the museum. The museum was a converted barn with a tin roof. The large barn doors had a ticket booth attached to them so that they were permanently closed. Over the ticket booth was a large sign with blue letters, accented in yellow, with the words, "World famous gangster museum. Meet Al Capone." The entire display carried the feeling of a circus sideshow. The loudspeaker over the ticket booth echoed "You dirty rat," and other corny gangster movie cliches.

Steve, Jackie, and Mike paid at the booth and walked into the museum. Inside the museum, the first room they entered had purple lights. The scene behind the glass was a nightclub. The inscription on the plaque stated, "Informal execution." The scene vividly showed a gunman spraying rifle fire at a table and the victims captured in startled and morbid poses.

Mike elbowed Steve in the ribs and joked as he pointed the camera at one of the victims, "That's you Buddy."

Steve pointed at a point on the wall, which didn't match the surrounding wall. He seriously stated, "No! That's me." Steve walked over to the different spot on the wall and pushed with his palm at random places in the area. He found nothing and turned around to walk back to Mike and Jackie. He had a disappointed look on his face and leaned against the wall with his hand. He heard a click and the mismatched part of the wall behind him moved slightly. Steve looked at his hand against the wall and turned around to look at the place where the wall opened. He pushed against the loose panel and walked into a small room. In that room was a toilette and sink. He walked out with a sad look on his face. He spoke with disappointment to Mike and Jackie as he held up a piece of toilette paper, "I found the big secret room."

Jackie replied, "Now that is important."

Mike spoke seriously, "That doesn't mean anything. This place is cheesy enough to hide something sinister. Nobody would look twice here. If I were the bad guy, I would put the computer in here. Keep looking."

Steve nodded. He walked ahead of his group, and randomly touched places on the walls. All three of them looked for electrical cords and boxes, which could hold a computer. The rooms were narrow and the passageway was almost like a maze. The rooms and hallways snaked back and forth. The lights alternated between red and purple. Sometimes the group came to a place where a crack in the wall allowed outside light in. They walked past pictures, pistols, period clothing, and wax figures. Mike captured the whole tour on tape with the camera.

A young couple met them and went past them. The two young people gave the group confused stares and chuckled as they went by. An old man met them and stared at them as he walked by. Each time new people went by, Steve stopped poking and prodding. He tried to appear to be interested in the display. A family passed them and Steve walked over to his two friends. He whispered near their ears, "I feel like an idiot. We have been in here for two hours and not found a thing."

The man from the ticket booth walked up behind them. He was a mature man, in his fifties, with wavy gray hair and a mustache. He saw Steve feeling for spots on the walls and asked, "What are you doing? Is there a problem here?"

Steve answered, "No sir. There isn't a problem."

The older man said, "You folks have been in here since noon. Maybe it's time you went home." He looked at Mike, "People don't usually videotape this place. Who are you people?"

Mike answered him, "Just tourists. We just love gangsters."

The ticket booth man frowned and spoke, "Nobody loves gangsters that much. I noticed you left the bathroom open. That room is not open to the public."

Steve answered, "Sorry sir. We didn't mean any harm." The man from the ticket booth looked closer at Steve, and mumbled, "You look familiar. Don't I know you?"

Steve responded quickly, "No sir. We have never met. I just have one of those familiar looking faces."

The ticket booth operator stood closer to Steve and frowned. He said, "No you don't! I think there is something going on here. Maybe I need to call the police. I'd like to see some identification."

Steve answered, "We don't have to show you ID. We're leaving now!"

The older man reached for Steve and grabbed him by the arm. He raised his voice, "This is my property and I have the right to. Now you just hold it."

Steve twisted away from the older man's grip and pushed him back with his other hand. He yelled, "Don't you touch me!" He faced the man with a ready stance. His fists were clenched and raised. His elbows were tucked in tightly by his side. His left foot led his right foot slightly.

His heartbeat echoed in his ears and he was ready to handle an attack from this other man. Steve's mind raced and he concentrated on the other man's movements. He watched the expression on the man's face and his shoulders to predict where the attack would come from. Mike stood next to Steve with his hands slightly raised and his right foot ahead of his left. He stood ready with his right side forward and his left side protected. Jackie moved behind Steve. Her eyes were fixed on the other man. She felt around in her purse for her pepper spray canister. She pulled it out and held the can in front of her.

The man from the ticket booth paused for a moment and thought about the situation. He could see that he was outnumbered and he lowered his hands. He talked in a low, angry voice, "I think I'd better call the police." He pointed at Steve and growled, "I could handle this with just you right now! Nobody comes into my place and tries something like that. I know there's something wrong with you." He reached into his pocket and pulled out a cell phone. He dialed up a number and began talking, "Hello police?"

Steve lunged forward and knocked the cell phone out of the man's hand. It fell to the ground and broke in half. The other man was off balance for a moment, but recovered and swung at Steve. Steve watched the punch coming. With his adrenaline rush, the punch almost seemed to be in slow motion. Steve remembered how to parry and punch from boxing as a boy on the playground, and reached up with his left hand. He pushed the punch away to his right. The other man's momentum kept him moving toward Steve.

Steve thrust his right fist upward into the man's stomach, right below the ribs. This spot was the man's solar plexus or breadbasket. The other man collapsed to the floor and gasped for air. Steve and his group backed out of the room, turned, and ran out of the museum.

Mike said to Steve as they both left the building, "Nice shot in the old breadbasket. He'll be down for about a minute. You could have stuck around longer and really messed him up."

Steve replied, slightly out of breath, "Sounds like fun, but I think I've committed enough crimes today."

Steve, Mike, and Jackie ran to the car and got in. Steve tried to start the car, but the battery was weak from the gadget. On the third try, he got the engine started. Steve took a last look at the museum and saw the ticket booth operator stagger out and move toward them.

The back tires lost traction for a moment, and Steve felt the tires dig into the dirt. The tires spit a small cloud of dirt backward. Steve eased pressure on the gas pedal and the right rear tire caught a rock. The forward force was just enough to free both tires and get the car moving forward and out of the hole. The car jumped violently out of the hole and fishtailed. Steve bounced the car down the dirt road away from the museum and toward the road.

Larry spoke up, "Do I need to ask what happened?" Jackie answered, "The owner decided we were trespassing."

Mike pulled the tape out of the camera and reached for Larry's camera. Larry interrupted him and turned the camera away, "No wait! I'm busy. The computer is on and he opened your file. I'm in! I'll be busy for a couple of minutes. This computer is slow."

Steve said, "What? That was it?" Larry shouted, "Shut up! I'm busy."

Everyone in the car was quiet. They pulled onto the main road and headed away from Lantern Peak. In half an hour, Larry broke the silence, "I rewrote his messenger and electronic mail software. Anything he sends or receives goes through me first. I messed with his e-mails, infected his computer, and placed a hit on him."

Steve sounded worried as he spoke, "What about the police? He said he was going to call the police. Aren't they going to be after us?"

Larry laughed and replied, "Not likely. The police would be in the way of his hit on you. He went right for your file, so he knows who you are. He thinks he sent three hitters after you, but I changed his e-mails before they left. I'm in his motor vehicle records, mortgage records, and credit cards. I took his personal information and built a file to send to the hitters. He thought he sent them your file, but he sent them his own. From now on, every time he gets on the computer, I'll be there."

Steve asked, "Do I have anything to worry about?" Larry said, "Yup!"

Steve gave a confused, "Why?"

Larry explained, "He is coming for you himself, and he is smarter than you are. He also knows this area better than you do."

Steve was annoyed and raised his voice, "He is a point of contact, not a hit man. He is an old man and way out of shape. I defeated him with my hands. He is no match for me."

Larry scolded Steve, "To be what he is, he had to be a hit man at one time. He was smart enough to succeed at being a hit man. He was even smart enough to retire, which very few hit men get to do. He doesn't have to defeat you with his hands. He can defeat you with his brain, and he is smarter than you are. I'll bet that right now, he is going to get his rifle from its hiding place. Don't get too cocky! This guy is a tough customer. He has had years of experience at killing people, and he has never failed at it. Do you still think he is an easy match?"

Steve asked, "Okay, what do I do now?"

Larry instructed him, "Stay far enough away to be out of sight, but close enough to watch. I'm putting a map on the screen. The next thing you want to do is hide this car. He saw you with it. Turn left up ahead and go to the spot on the map."

An hour later, they hid the car near some trees about a mile south of the museum. From their hiding place, they could see the museum and nearby house. Trees and brush provided cover between them and the museum. The landscape, though very flat, had mounds and small hills. A ditch ran near their position and ended near the museum. Larry instructed Steve and Mike to use the ditch to get in close and videotape the area, but not to leave the ditch or stay very long. Larry watched the target from the car with his telephoto lens. The museum was closed and there was no vehicle nearby. Steve and Mike returned to the Beast and Larry.

Larry watched Steve and Mike's tape.

Steve asked Larry, "Since there is no car there, can't we go inside and wait in ambush?"

Larry returned a question, "Do you know how to break in? Just because there is no car there, are you sure he isn't there? How good are you at ambushing anyone? Can you get near the house without leaving tracks? We don't want to approach the house at all. We want to stay out of the way of the hitters and let them do the job. You are only safe, once you know the hit has been made. We are just here to watch."

Steve was quiet and everyone waited. Steve started the car every hour and ran it for fifteen minutes to keep the battery charged. Once it got dark, Jackie covered the computer screen with a blanket to hide the light. She only looked at the screen when Mike told her it was safe, and then only for a few minutes.

Shortly after ten in the evening, a pickup truck arrived and parked near the museum. The man got out, opened the booth of the museum, and turned on the light in the booth.

Larry spoke, "He is on the computer again. He is instant messaging the hitters to find out where they are."

Steve asked, "Anything else?"

Larry answered quickly, "Quiet. I'm waiting for the answer." Everyone was quiet. Fifteen minutes later, Larry broke the silence, "The hitters are in place. I changed the instant messages. He thinks the hitters are at the hotel you were last night. I see one of them at the end of the ditch. Don't move or make a sound." A few minutes later, the light in the ticket booth went out and a shot echoed in the dark. Three figures approached a figure on the ground. There was another shot and the figures left. Car headlights appeared in three places. The lights moved away. Once the lights disappeared, Larry spoke again, "We can leave now. The job is done."

The Beast started and pulled away. An hour later, Larry announced, "The hit men checked in and I paid them. I closed the network and Internet account."

Steve replied, "That's great! It's late though and I am going to pull into that motel over there. We all need sleep."

Larry asked, "What about me?"

Steve said, "We'll shut you down and you can cruise the Internet until we get back. Don't get into any trouble out there."

Jackie clicked on shut down computer on the keyboard. The computer turned off and she pulled the plug from the cigarette lighter. They pulled into a motel and got three rooms. The next morning, Steve called in sick, and then went back to bed again. Shortly before noon they departed and started their return home. They returned to the hospital late that night. Mike and Jackie went straight to their cars and drove home. Steve went home too.

CHAPTER 8

THE PROJECT WAS SOLD

The next morning, all three were at work on time. The display on the large screen showed eighty four percent. Dave was there before anyone else arrived and staring at the screen.

Dave walked up to Steve and announced, "I made a big sale yesterday. Barron Motor Corporation bought a set of software. They brought a small team of neurologists as consultants. They said that this was more than enough to serve their needs and they paid nineteen million dollars for it."

Steve asked, "What are they going to do with it?"

Dave replied, "They said something about cars that can read road signs, don't drink and drive, and don't get distracted while driving. They even said that they could change laws and wipe out the competition if they could bring a prototype to market first."

Steve said, "They told you all of this? Aren't they worried about proprietary information and patents? Aren't they worried about the other companies doing the same thing?"

Dave boasted, "They made me sign a stack of contracts which prohibit the sale of this software to their competitors and give them the exclusive right to automotive applications of this product. If competitors get this from someone else, they can sue the seller for millions. We have to attach a clause to each software set we sell to inform buyers of the restriction."

Steve was suddenly uncomfortable and worried. He asked Dave, "Did they seem threatening or anything? When did they leave? Has anyone been following you?"

Dave broke out in laughter and answered Steve, "You've been watching too many spy movies lately. They weren't threatening. They even told us that they would help us market this product outside of the automotive market. Don't worry Steve, we have counseling available on your insurance." Dave patted Steve on the shoulder and chuckled as he spoke, "It's okay James Bond. Quartermaster branch has this one covered." He walked out of the door of the lab and back toward his office.

Steve walked over to Mike and Jackie. He whispered, "Dave Anderson sold half of the rights to this brain to Barron Motor Corporation."

Jackie smiled and spoke, "That's great! We are in the dough."

Mike whispered, "Why are you whispering?"

Steve whispered louder, "We may be in the dough, but we are in trouble too! Anyone see the movie, "Tucker?"

Those guys played dirty fifty years ago. Imagine what they are like now?"

Jackie spoke to calm Steve, "They played dirty because Tucker wasn't playing ball with them. We are. Barron isn't mad at us."

Steve moved closer and whispered quietly, "How do you think Ford, Chrysler, and GM feel?"

Mike and Jackie were suddenly quiet. Mike walked away and came back a few minutes later. He grabbed Steve by the sleeve and pulled him closer. He whispered, "What is next? What do we do? Do you have a plan?"

Steve pulled his sleeve away from Mike and spoke up, "No I don't! I don't know if anything is going to happen. I don't know if it got worse or better. I don't know if we should do anything at all. I don't know anything at all! That's what worries me!"

Over a thousand miles away, in a Barron Motor Corporation central office, there was a meeting in an office. Twelve men sat around a large table. The man at the end of the table waited to hear reports from his subordinates. The mood was somber. A small man near the other end of the table stood up and addressed the senior man at the table, "Sir, we have the entire software product. They are bound by contracts. We can go ahead with the project."

The man at the end of the table stood up. He was a tall man. He was thin, but muscular. He had gray hair, but plenty of it. In this small group, he was the meat eater, not the prey. He addressed the group, "Well then, we don't have to worry about anything. If they do anything stupid, then we will own Gm, Ford, Toyota, and everyone else. Bill Gates thought he had a monopoly!" He pointed at one corner of the table. He grinned as he spoke, "Take the South Building! Make it the new research branch! Get scientists, lots more computer engineers, get anyone and everyone to make this thing real! Get Pentium to make something for us." He turned and addressed another man on the other side of the table, "You! Get loans for research!" he pointed at another man two seats down from that man. He spoke loudly with a menacing edge in his voice, "You put together a proposal which makes his banks salivate. I want them begging!" He walked around the table to a man on the other side of the table. He placed his hand gently on the man's shoulder. He spoke in a normal voice, but had a bullying edge to it, "You will build the smartest car in the world. It will be a better driver than anyone else. If you don't, I will personally eat you for lunch! You have three weeks!" He patted the man on the shoulder twice and moved two seats down. He whispered loudly enough that everyone at the table heard him, "You will put it all together! The other men will have funding, buildings, equipment, and fear! If they don't, I will use your head as a paperweight. Make no mistake. I mean it!" The boss took a relaxed stance and looked around at the entire table and paused to look each man in the eyes. His next words were low and soft, "You are all still here?"

The room emptied briskly with the sound of, "Yes sir!" Eleven people scrambled for the door. Within two hours, trucks with mobile home trailers entered the main gate of the complex. Barbed wire fences were erected, and building permits appeared at the south end of the area. Trucks with computer equipment arrived, and construction companies lined up outside of the main gate. The boss, who had just led the meeting two hours ago, looked out of his office window. He glanced at his secretary and said, "I finally have the big edge I have waited for! I own the industry! Ten years ago, we struggled with Bankruptcy! Now we have it all! Get my lawyers! I want to buy that hospital!"

Back at the hospital, Dave came into Steve's office and spoke to him in a worried tone of voice, "Steve, there is something going on. Barron submitted a bid to buy the hospital. Maybe you were right. Any ideas?"

Steve looked back at him and replied, "Then everything is fine. If we are employees, then they won't want us dead. They will approve all sales of the product and won't have any reason to do any more to us. We aren't in trouble anymore. We will do what they say, and they won't be mad at us. Ford, Chrysler, and Chevrolet won't mess with us because we are part of Barron. That is good news. Thanks boss. I'm not worried anymore." Dave walked out of Steve's small office with a worried and confused look on his face. Steve stood up and walked to the lab. He found Jackie and Mike. He whispered with an excited tone of voice, "We are in the clear! We are now part of Barron Motors. They aren't out to get us anymore."

Mike was still worried and showed it in his voice, "What about the other companies?"

Steve smiled and spoke confidently, "Barron will protect us. We belong to them. The other guys can't mess with us now."

The large computer screen showed eighty seven percent complete. Larry spoke over the speakers, "Steve, we need to talk! Would you get a microphone so we can do a better job?"

Steve typed, "Sure, I'll get mine, go to supply, or go to the store."

Dave appeared behind Steve and spoke to him, "Grumman is on the line with me. They want to talk."

Steve turned around and faced him, "Then you need to call Barron and ask if it's okay. Why are you asking me anyway?"

Dave showed anger in his voice, "Because you got me into this with your project. Grumman does a lot of government projects. That gets us in deep."

Steve grinned and waved his boss off, "Then take two Alka-Seltzers and call me in the morning. All we need now is IBM. Oh, that's right! Barron is buying the hospital and they will get all of the money from the sale of this product. What was I thinking?" Steve chuckled, "They will get their money back from us. The profits from the project will reimburse them the cost of buying the hospital. After the project is finished, they can dump the hospital again and pocket the difference." Dave stormed off shaking his head.

Larry spoke up again, "Steve, the microphone?"

Steve replied, "Right! I got distracted. I'll be back." He went to his office and pulled a microphone out of his computer and walked back with the plug. He asked after he plugged the microphone into a terminal, "Don't you need the voice recognition software? I don't know if this will work."

Larry answered, "It works fine. I don't need voice recognition software. I am not a computer. I remember what it's like to hear. I didn't need it in your car did I?"

Steve asked, "What did you want to talk about?"

Larry spoke, "While you were gone yesterday, your people were busy watching snuff films from my memory. Dave hovered around here constantly, so I couldn't say a word. I saw some e-mail traffic from the CIA. I have a feeling that the project will be shut down in a couple of days. If they shut down this computer, I'm dead. I need that place to stay that we talked about."

Steve answered, "That's right! I need to get a computer for you."

Larry whispered over the speakers, "Yes, but here's what you need to do. Load the writing program into it. Load the map of my brain, and then I'll download what is in the RAM here to it. The money I sent you should be in your mailbox soon. That will help reimburse you for the computer."

Steve thought for a moment and then asked, "Can you show me the e-mails from the CIA?"

Larry answered, "No, but they will be here today. They are more interested in the technique than the results."

Steve laughed as he spoke, "They are out of luck. That part of the project is over. The team has been dismantled.

The slicing equipment is gone. I didn't record the technique on video or anything. I can't give them what I don't have."

Jackie chimed in, "They missed the show by about two weeks. We don't have another brain anyway."

A tall black man in his forties spoke from behind Jackie, "You don't need one. We brought one."

Steve turned to face him and said, "Well Sir, I'm assuming you are with the CIA. I guess I'd better take a good look at your badge now, because I have a feeling that I will wish I had later."

The man smiled and in one motion, pulled a leather pouch from his breast pocket, flipped it open and laid it in Steve's hand. He spoke calmly, "Sir, Agent Brian Stills at your service."

Steve grinned and remarked, "I believe the service roles may be reversed. I have a feeling that I will be in your service instead." Steve returned the agent's badge. Steve spoke again, "I have to be honest. I don't know how to tell a real badge from a forgery. How can I know for sure?"

Brian explained, "I guess that the best proof is that I have been keeping track of your project. I followed your little caper in Nebraska. I also know that you are establishing your innocence with Detective Edward Morris. Your wife stepped out on you and put a contract on your life, and you have twelve messages on your answering machine." Brian lifted a picture from a computer station. Wires and a small microphone were behind it. He added, "Do you need any more proof?"

Steve nodded and spoke, "Okay, I believe you. You might as well tell me now what is going to happen next. You seem to have a plan in mind."

Ten men entered the room wearing orange coveralls and carrying boxes. They began setting up video equipment all over the room. The large slicing table returned to the corner where it had been before. The refrigerating unit also returned. Brian put his hand over Steve's shoulder gently, but Steve felt very uneasy. Brian explained to Steve, "Doctor, we have watched your results. We want to be able to pull information from a subject's brain. If the subjects are dead, which they often are, we would like to be able to debrief them. We need to duplicate your technique." Brian led Steve over to a computer station and placed a briefcase on it. He opened it up. It was full of hundred dollar bills. He patted Steve on the shoulder and spoke patronizingly, "Eighteen million dollars will reimburse the hospital for the cost of the operation and give everyone a healthy raise. What is in this briefcase is your four million dollars. My people will take samples and study your method. I already talked to Doctor Anderson and told him that we are buying this project and the equipment. I told him that we would be repeating the experiment with new security procedures. Go ahead and bring back anyone you need to do the job. My people can fill in some of the slots too. The whole operation is now the property of the CIA."

Steve closed the briefcase and took it. He knew that he didn't have a choice, but to do what these men wanted. He went to a phone and called Doctor Jordan, "Jim, would you consider doing this job again for four hundred thousand dollars?" The answer was loud enough that it hurt Steve's ear, "I'll be there tomorrow!"

He walked back into the lab and saw Brian's group huddled around a set of computer stations in the corner of the room. They were marking spots on a large printout of a brain, and writing notes in notebooks. Four of the men were reading memories from Larry's brain. The men were whispering to each other. One of the men was printing neural pathways at a printer. That corner of the room was alive with the sound of hard drives, printers, and CD drives. Two of Brian's men were taking one of the scanning stations apart and putting it in crates. One of Brian's men stood by the entrance door to the lab with his arms crossed and his back to the wall. Steve could see part of a leather strap under his jacket, so he knew that this man was armed.

Steve approached the group in the corner and spoke to Brian, "I called the man who runs that machine. He will be here tomorrow." He pointed at the large slicing machine."

Brian glanced at Steve, nodded, and spoke, "Good. Thank you. Will that be all?" Steve answered, "Yes."

Brian hurried him away by saying, "Then thank you and please excuse us."

Steve walked away from the commotion in that corner. He went to the bank and deposited his new paycheck. When he returned, he called Ed, "Yes, detective. The project isn't finished quite yet, but I believe that there is enough recorded to keep you busy for weeks. I think you'd better come down here and collect it before the project gets shut down."

Ed asked, "Shut down?"

Steve quietly spoke, "An outside agency. I can't talk."

Ed answered, "I'll be there." He hung up.

Steve returned to the lab. Another part of the room had several operators compiling disk sets of the software for the customers. These were Steve's employees, and they were making the product for their original customers. Steve collected one set for the detective. He placed one hundred twenty disks into a CD case, and slipped it into his breast pocket. When Ed arrived, Steve handed the case to him and said, "This is eighty five percent of the man's brain. I believe that's enough." Ed looked over in the new busy corner of the room with a puzzled look on his face. Steve patted him on the arm and whispered, "CIA, don't ask. You had better leave now before they become interested in you."

Ed said, "Thank you. I thought you might like to know that we are still holding your attacker and wife." Ed walked out. On his way out of the lab, he glanced at the big screen. It showed eighty seven percent complete.

Brian left his think tank corner and walked up to Steve. He spoke, "We have enough information from this brain and we don't need to collect the rest of it. We can pull the plug on the old project now and move on to the new brain."

Steve answered, "No you can't. The man who does the first part of the project won't be here until tomorrow. Even then you won't have enough to do anything with until another day or two of slicing. In the meantime, we can continue to harvest this brain. It would be a real waste of time and effort to scrub the project now, and nobody would gain anything from it. Once you have a few slides scanned into the computer, you can save them into another file and still not need to interfere with our operation. You will get everything you want without stopping this project."

Brian looked frustrated, but agreed, "All right. You can keep going. You'll be done soon anyway. We need the memory space cleared soon though."

Steve remarked, "There is plenty of memory space for our project and yours."

Brian turned around and gathered his group from the corner. He led them to a conference room a few doors away from the lab. Steve went to his office, which was across the hall from that conference room. He tried to look busy as he tried to listen in.

One of the scientists stood up at the end of the table and pointed to a spot on the big drawing of the brain. He described it, "This spot right here is like an interstate. Messages go back and forth. Most of the new things, which happened, are right here. If we build a microchip, which can interface right here, we can retrieve data from this spot. This one small spot will give us weeks of memories for the asking. We can surgically implant the chip here in a live subject and read all of his recent thoughts. The subject won't suffer any damage, and the scar will heal quickly. The subject brain we brought is only of interest in this area. We don't need the whole thing. I don't care about his feelings or childhood. Do you?" Everyone in the room started laughing.

Brian got up and addressed his group, "That's enough for today. We can't do anything else until tomorrow anyway. Gather up your notes. Don't leave anything lying around for the civilians." Brian paused and turned to one of his men. He gave him an order, "Design the chip and get ten made." The other man nodded, the conference room emptied. They all left to go to their motel. Brian glanced in Steve's office as he left. Steve was looking down at some papers.

When Steve heard the CIA vehicles leave the parking lot, he returned to the lab and found Mike and Jackie. He pulled them away from the computer terminals and whispered, "Watch what you say and do around these guys. I have a bad feeling about them. They are looking for new and creative ways to interrogate victims."

Mike asked, "Victims?"

Steve answered, "That's what I call anyone they get a hold of. Do you want to be one? Just watch it."

Everyone finished their work and went home. The computer continued building its model. Steve was the last one out of the lab and he turned off the lights. The last thing he could see in the room was the big screen. It showed eighty nine percent. He heard a good night from the room and he returned it. He locked the door.

The next morning, Steve met Jim at the lab door waiting to get in. Jim said, "Finally showed up huh?" Jim had a gallon jug of liquid in his hand.

Steve answered, "Yes, I did. Watch what you say around these guys they are with the CIA. They have doctorates in paranoia, and they are armed."

Jim asked, "CIA? Figures. Don't worry; I'll put them to sleep with my machine."

Steve replied, "Gee. I hope so."

Steve and Jim just finished pouring their coffee when the entire government group marched in. Brian walked right up to Jim and held his hand out to him. He greeted him enthusiastically, "Doctor Jordan. I am Agent Brian Stills. You are the star of the show today. We are very anxious to watch you work on our new subject."

Jim took Brian's hand and shook it. He joked, "It wouldn't happen to be my ex wife would it?"

Brian smiled and responded, "Not unless she had a sex change."

Jim said, "Too bad. I never could figure her out." Everyone in the room heard it and laughed. Jim led his group over to his machine and said, "The subject has to soak in this for twelve hours and then freeze for eight hours." He poured the liquid into a stainless steel pan and waited for the test subject.

One of the government men put a box on the table and removed a brain from it. Another man cut out a piece from the middle of the brain and placed it into the liquid. Brian asked Jim "Isn't there any way of speeding this part up?"

Jim replied, "Not if you want the slides to be readable. If the fluid doesn't get into all of the pores, the cells will move when I slice it. The subject needs to be hard as a rock for this to work."

Brian looked at two of his men and nodded. They gathered up the sample and placed a lid on it. He spoke to Jim, "We will be back at six tonight. You'll need to freeze it then." He led his group out of the lab.

Jim waited until the group left and whispered to Steve, "I just got my orders. I guess I'll go down to maintenance, build a carrier for this thing and start the freezing tank."

Steve nodded. Fifteen minutes later, Mike, Jackie, and the rest of Steve's group arrived. The display on the big screen showed ninety six percent. Larry spoke to Steve, "Steve, do you have the computer?"

Steve replied, No, I don't. I'll go do that now." He turned to Jackie and asked, "Would you come with me and help? Don't forget the writing program and the map."

Jackie gathered up discs and followed Steve. Jackie spent three hours getting the computer hooked up, online, and full of the new software. They returned to the lab after ten. Steve went to the terminal with the speaker, and spoke to Larry, "I'm typing in the e-mail address for you. It is online and ready for you."

Larry answered, "Thanks Steve. I'll see you when I get to my new home." The display on the big screen showed ninety nine percent. Larry spoke over the speakers again, "I can't get in. Something isn't right with the software. The writing program isn't running. I can look around in the computer, but I can't leave here. Steve, you need to get that fixed. I don't have much." The voice stopped. The display on the big screen showed one hundred percent. The screen went blank. The writing program shut down. The picture of the brain disappeared. The room became quieter.

Steve spoke up, "Larry?"

There was no answer. Jackie asked, "Larry, are you there?" She went over to a terminal and typed, "Larry?" Nothing happened. Mike and Steve looked at Jackie and shrugged their shoulders. Jackie solemnly said, "He's dead. I couldn't save his life."

Steve could see tears welling up in her eyes. He went over to her and held her to his chest. She held him close. Mike walked away and was very quiet. He shook his head as he walked away. Steve spoke softly to Jackie, "I think everything is fine now. Larry did a good deed and I think his soul finally went to heaven. Do you remember what we saw when we first read his thoughts? Remember the visions? Remember?"

Jackie backed away and looked in his eyes. She smiled and nodded her head. She wiped her face with a handkerchief and nodded again. She spoke, "I think so. I think he made it to heaven this time. He was alive in all of those terminals because the writing program was testing new connections. Once the program finished mapping out the brain, there was nothing left to write or test. The program just shut him down." She smiled and walked over to a computer terminal. She sat down and stared at the blank screen. She spoke to the screen, "Sorry Larry. You didn't have much time for your second life. Thank you for your help though." Steve left and went home to check the terminal. When he got there, the screen was blank. Larry didn't make it to his new home.

When Steve returned to the lab, everyone was busy making copies of the map. CD writers slid open and shut all over the room. CD cases full of software piled up in a corner. Steve counted forty-three sets of product. He asked Mike, "How many sets are we making?"

Mike answered, "One-hundred ten. One hundred for customers and another ten so we can have it too. Call it a fringe benefit. We all poured everything into it and we all want to remember Larry."

Steve frowned and told Mike, "I sure don't want anyone going into business for themselves."

Mike replied, "You'd better tell that to the detective.

Don't worry. I put a copy guard on each of these. They can't be copied or changed. Jackie built the copy guard into the software set."

Steve nodded his head and found a terminal. He started copying sets too. In a couple of hours Jim returned and shortly afterward, so did the CIA. Jim placed the sample into the freezer while Brian and his men watched. Once the sample was in the freezer, Brian posted three guards. Jim shook his head as he walked out of the lab. An hour later all of the copies were done and stored in Steve's office.

Mike handed Steve a red CD case. Mike said, "This set does not have a copy guard. It is the master set. You'd better put it in a safe place. Jackie is wiping out the map on the system. This is the only way to restore it." Steve took the pouch and hid it in his office. Everyone slowly filtered out of the room until the only ones left in the lab were the guards.

The next day, the lab became a busy place. Jim placed the new brain sample into a carriage and mounted it to the slicing machine. As Jim turned the machine on and began to make slides, he had a group of ten CIA agents around him watching and taking notes. Two of the agents busily filmed the operation. After a few minutes, they took over and began to do the operation themselves. Very soon, Jim was standing away from the table and couldn't even get close enough to see what they were doing. He did remind them that the sample had to be returned to the freezing tank every three hours to become rigid again. When it was time to return the sample to the freezer, Jim stood back and told them what to do.

Brian walked up behind Jim and told him, "Thank you Doctor Jordan. Your work is finished here. You have done a great service for your country."

Jim backed away and answered in a surprised tone of voice, "Your welcome." Jim found Steve and said, "I've been sent on my way, so I guess I'd better collect my paycheck now."

Steve nodded and replied, "I'll be back with it in a minute."

Jim tugged on Steve's sleeve and whispered, "Something isn't right with these guys. I don't think they want me to see what they are doing. They are hurrying me out of here. You'd better watch yourself."

Steve nodded his head and said, "Thank you." He took care of Jim and his paycheck.

Jackie and two of her helpers began to scan the slides into the computer. Ten agents crowded around her. They were watching, filming, and taking notes. After a few short minutes, they took over, and Jackie couldn't even get close to the scanner. Jackie went to her computer station and pulled out the software for reading thoughts and writing the map. She placed it into the CD carriage of her terminal and the big screen turned on. The agents looked around in confusion when she did this.

Brian walked up behind her and spoke softly, "What is that?"

Jackie looked up from her terminal and found four agents around her filming and taking notes. She looked at Brian, held up software disks, and spoke, "These disks have the programs on them which trace the neural pathways and the programs which translate the patterns into something which a computer can read. This software will do all of the work of making a computer model of the brain and then reading the mind."

Brian looked down, smiled, and then took her hand. He pulled her away from her terminal. He whispered, "Thank you very much for your contribution. We will take it from here."

Jackie resisted and tried to get back to her disks, which were being returned to sleeves in a carrier. One of the agents tucked the carrier into his pocket and walked away. Jackie tried to follow, but Brian held her hand and stopped her. He spoke a little more loudly, but still softly, "Don't worry about anything Miss. We will take it from here. Your software is safe with us."

Jackie spoke up, "That is my software. I want it back. What are you people doing?"

Brian's voice was firm, "We are doing our job. We must have complete control of this operation. You have done your job and may go now. We are keeping the software. Thank you Miss."

Jackie turned around and scowled at Brian. She pulled her hand free. She spoke, "You people are a real pleasure to deal with!" She walked away shaking her head. She found Steve and told him, "I've been dismissed! I guess I'd better leave now before I get shot. They will probably kick you out of here soon too."

Brian walked up behind Steve and whispered over Steve's shoulder, "You are both dismissed now. We will take the project from here. This is sensitive information and you don't have the necessary clearance. These men will show you out." Two men stepped forward behind Brian and motioned Steve and Jackie toward the door of the laboratory. Steve and Jackie left as they were told. Brian stood and watched them leave. He had a subtle grin on his face. Once Steve and Jackie were out of sight, he spoke to his crew, "Pack it all up people! I want everything on the trucks before we leave here. You can leave the coffee maker this time. I want the slicing machine, computers, and all of the software. Let's go! The clock is ticking." The entire crew went around the room turning off computers. They opened toolboxes and began moving equipment around the room.

One of Brian's men whispered to him, "Sir, would you like for me to take care of the witnesses?"

Brian smiled and said, "I'm not worried about them. I may need them again anyway. I think I'll just keep an eye on the good doctor and see what he makes for me next."

CHAPTER 9

THE BAR

Steve and Jackie met Mike at the parking lot. Mike spoke up, "I guess they needed both of you a little longer than they needed me."

Steve was annoyed as he spoke, "Yeah! Just long enough to get what they wanted. I wonder what will be left of the lab when we get back."

Jim pulled up on his motorcycle, and smiled as he spoke, "They gave you the heave ho too?"

Steve replied, "It was a need to know thing and keeping the world safe stuff."

Jim answered, "I boxed up the software discs and put them on your car while they were busy kicking you out. I even saved the red disc carrier for you." He pointed at a set of boxes on top of a Lexus.

Steve spoke up, "That one is Mike's car."

Jim asked, "Okay, which one is yours?" Steve pointed at the Beast. Jim leaned back in his seat and laughed. "That's good Steve. Did you borrow it from your son? I guess the BMW is in the shop."

Steve cracked a cynical grin. He replied, "Yeah, that's what it is. Speaking of teenage vehicles." He pointed at Jim's Harley Davidson. Steve walked over to the boxes of CD's and put them in his own trunk.

Jim smiled and spoke, "My kids and grandkids drive BMW cars. Man, are they uptight! I do what I want and drive what I want. When they tell me to grow up, I ask them what good it did for them. I don't know about you proper medical and computer professionals, but I feel like being a kid for the rest of the afternoon. I'm going to a bar. I might even start a fight. Anyone joining me?"

Mike replied, "Drinking heavily and Being rude, I like it."

Steve smiled and chimed in, "Don't leave me out. I wouldn't mind picking a fight with someone who looks like that agent Brian, besides, I can't go back to the lab. I don't work there anymore. I don't want to go home. It isn't much fun anymore. I'd rather be with people I like."

Jackie came up behind Steve and held his hand. She spoke softly as she squeezed his hand, "Me too."

Steve growled, "What really burns me up about this is that they took my project away from me. I'll bet that they are taking all of the equipment too. I can't stop them, because they bought it. I feel so doggoned violated!"

Dave walked up behind Steve and put his hand on Steve's shoulder. He spoke to the group, "Are you all going to leave early and do something irresponsible?"

Jim laughed and replied, "Yes, and you are our designated driver."

Dave pointed at an approaching taxicab and said, "Oh, no I'm not. I need a beer to cry in. That's our designated driver. Don't ask me about my day."

Steve was puzzled and asked, "Are you going to fraternize with the troops?"

Dave answered, "Yes, I am. Do you want to do something about it?"

Jim put his arm over Dave's shoulder and spoke softly, "Just keep that attitude, and your day will improve." Jim turned around and went to his Bike. He climbed aboard and before he started it, he asked, "Coming?" Steve, Jackie, Dave, and Mike squeezed into the back seat of the cab. Both vehicles left the hospital area and went East on Colfax Avenue. Fifteen minutes later, they arrived in front of a bar.

This bar looked like a small version of a barn. The paint was cracked and there were holes in the front door.

Everyone paused in front of the establishment except Jim. Jim walked up to the front door and grabbed the doorknob. He looked at his group of guests. He smiled and said, "This is home." Everyone was quiet. He pulled the door open and before he went in, he said, "If you want to hang around the BMW crowd, they are that way!" He pointed toward the mountains, then walked into the bar.

Steve looked at the rest of the quiet group and shrugged his shoulders. He smirked and said, "You saw my car. I'm not part of the BMW crowd. Any body for a good fight?" Steve disappeared quickly into the bar.

Dave laughed and replied, "I'm on my way!" He walked up to the door and followed Steve in. Mike and Jackie followed. Dave walked into the room behind Steve and held his fists clenched. When he entered the bar, he found Jim surrounded by a group of twelve friends. Jim introduced his group to the bar, and before Steve's group could say anything, Jim's friends made the new group feel at home. The bikers bought Steve, Jackie, Mike, and Dave pitchers of beer.

Jackie pulled Jim aside and asked him, "What is wrong with the BMW crowd?"

Jim replied, "Nothing if you like to be stylish and a trendsetter. There's nothing wrong with having a nice car, nice job, and nice bars. It just doesn't have very much to do with me. I like this place because it's friendly. When I go to the other places, I feel alone. This is home."

Mike walked up to Jim and asked, "Where is the Lexus crowd meeting?"

Jim smiled, patted Mike's shoulder, and said, "They moved their meeting right here." Jim pointed at one of his friends with a very short haircut and said, "He has two. A red Lexus and a green one."

Jackie began telling the story of the adventure they had lived for the last month, and the whole group crowded around them.

The biggest of the bikers stood up and spoke in the middle of Jackie's story, "I'd like to meet this rain man. I like the duct tape part. Too bad I missed that."

The female bartender interrupted him and chimed in, "Slim, knock it off! We all know what you would have done to him. I just don't think he was ready for marriage yet."

Slim smirked and shouted, "There went your tip Sally!" The bartender replied, "You mean the whole dime?"

Everyone in the room was quiet until the bartender spoke up, "Sorry Jackie. I love your story. I just had to shut knucklehead up." Jackie blushed and continued. Slim put his hand over his mouth and looked at Jackie.

A black man entered the room. He looked very much like Brian. He was missing a patch of skin on his face, which included his ear, and his arm was missing on the same side. Everyone in the room turned around and looked at him. They all greeted him, "Hey telephone pole!" Jim grabbed Steve by the shoulder and looked him in the eyes, "Hey Steve. This guy looks like that agent. Can you take him?"

Steve blushed and replied, "If I could, then I wouldn't feel too good about it. If I couldn't, then I'd feel worse. What happened to him?"

Jim answered, "A telephone pole!"

Steve answered, "I'd rather just leave him alone."

Jim smiled big and patted Steve on the shoulder, "Son, I have learned that revenge doesn't taste sweet at all. Don't worry. It just happens. I think they call it Karma."

Steve smirked and replied, "I am humble in your presence, oh Great One." He put his hands together as if in prayer and bowed his head.

Jim remarked, "Okay wise guy, you can buy this round.

Last time I try to be paternal."

Steve frowned and spoke, "Is that any way to talk to your boss?"

Jim burst out laughing and yelled, "Check your calendar pal! We are all unemployed! Right now, you are the boss of that mug of beer! Don't let it tell you what to do! It is telling you to go to the bathroom. Don't do it! You can't follow orders from a mug of beer, can you?" Jim paused and smiled. He whispered, "I'll bet you are feeling it now, aren't you?"

Steve smirked and replied, "Thanks for pointing it out." Steve looked at the waitress and spoke up, "A round for the house on me!" He held up his credit card and handed it to her. He stood up and walked toward the sign for the bathroom. Under the sign was a small hallway. On the right side of the hallway were the restrooms. The men's room was close to the exit at the end of the hallway. On the left side of the hallway were a couple of dark rooms and the kitchen. Steve did what he needed to in the restroom and came back out. As he moved back toward the bar area, he felt a hand on his right shoulder and a sharp instrument at his lower left back. He stopped and was quiet. He heard a whisper behind him.

The voice behind him said, "Remember me?"

Steve replied, "If this is Dave, then it's your turn to buy the next round."

The voice answered, "Do you remember a Mister Richard Turner in a grocery store who lives on Elm Street?"

Steve chuckled, "Dick Turner. You should have picked a better name. A last name of Head would have been better for you." The man behind him pushed the knife harder into his back and it penetrated Steve's skin. Steve exclaimed, "Ouch you jerk! How would you like that frontal lobotomy I promised you?" Steve felt his own blood drip down his back. It was warm and gathered at his belt line.

Mister Turner whispered closer to Steve's ear, "You aren't in any position to do anything right now. I can have my way with you. I was kind of thinking of giving you a loboto my instead. I wasn't going to go in through the nose though. I want to take the long way and go through your back."

Steve was in pain and thinking very fast. He was also tense from fear and anger. He remembered how to roll off of a knife from a self-defense class he took, and knew that a blow from an elbow delivered more force than a punch. He realized that rolling off of the knife was very risky, but he would be dead in three minutes if he didn't. He controlled his trembling as he whispered back, "Dick, that's way too much work. You would waste way too much time, and you would never get the blood out of your clothes. Come closer and let me tell you how to really do it right." He heard breathing right next to his right ear, so he knew where the other man's head was.

In one motion, Steve shifted his weight to his left foot, lifted his right elbow, and spun his body around to the right. The knife dug in slightly and then rolled off. As he turned around, his sudden and unexpected movement caused his enemy to become off balance and push the knife forward into empty space. Steve lifted his elbow to head level and snapped it back as hard as he could. He connected hard with the right side of the other man's face. The impact was enough to hurl Mister "Rain Man" Turner against the wall and a light fixture, which caught him in the back of the head. Steve's inept assassin collapsed to the floor and the knife rolled gently out of his hand to the floor.

Steve reached down and found the back of the man's collar. As he established his grip on the man's collar, he spoke to him, "Consider yourself lobotomized, Dick." He dragged the man across the floor into the main area of the bar and spoke to the group, "Hey everybody! This is the rain man! Don't touch the knife in the hallway. It's evidence! I'm going to call the police." Four of the bikers stood up from the table and slid out of the front door as quietly as ghosts. Steve paused for a moment as he watched them disappear.

He spoke, "Gee, is that a problem?"

Slim stood up and walked over to Steve. He took the man's collar from Steve's hand. He dragged the rain man to one of the empty chairs left by a fleeing biker and placed him in it in an upright position. He held the rain man in place by the collar and spoke softly to Telephone Pole, "Would you get this man's duct tape out of his truck? It should be a black Dodge Ram." Telephone Pole stood up and smiled. He didn't speak. He just left. Slim looked at Steve and told him, "Steve. Those guys left because they have warrants out on them. Everyone here is clean right now. I am going to unlawfully detain this guy for a little while. I make tattoos in my spare time and have something in mind for him. Don't ask, just watch."

Steve asked Slim, "Why did you ask a disabled man to do that?"

Slim spoke seriously, "Because he'd be mad at me if I didn't. He doesn't consider himself to be disabled."

The bartender noticed Steve's injury and rushed over to him. She lifted the back of his shirt and stuffed a napkin in against the cut. She told him, "Don't move around so much. You are making a big mess. It's a real beauty. I think you are going to need stitches. Nice back though." She caressed the uninjured side of his back.

Jackie rushed up behind Steve almost instinctively, and replaced the bartender as Steve's nurse. She told her, "Thank you for the help, but I really should be doing this." She put her head close to Steve's and whispered, "I'll go get a needle and thread." She admired his back. This was the first time she had seen him without a shirt covering him.

Steve answered, "Okay, but I'll need to finish my beer to be sedated." Jackie smiled and slapped him on the shoulder.

Telephone pole appeared in the doorway and tossed the duct tape to Slim. Slim fastened the rain man securely in place against the chair and took his tattoo gun from his pocket. He wrote across the man's cheeks, "I want a man to love." Everyone in the bar broke out laughing. Slim spoke up, "If he wants to press charges, I'll be his cell mate. He'll like me!" Two of the bikers fell out of their chairs and began rolling on the floor laughing.

Steve asked Slim, "Is it okay if I call the police now?"

Slim chuckled and replied, "Oh, please do. I can't wait. I hate sneaky back stabbing jerks like this. They hide. They wait. They kill. It's not very manly or honorable. Usually the victim doesn't even know it's coming. A friend of mine went that way. Maybe it was this guy. This one's on me!"

Steve went to the phone to call Ed at the police department.

Sally giggled, "Slim, sounds like you are going to have a good night."

Slim growled, "Better than his!"

Steve returned and announced, "They are on the way now."

The rain man regained consciousness. He yelled as he woke up, "Hey! What is going on? My face hurts! What did you do?"

Steve leaned down and spoke close to his face, "You and I talked, then Slim wanted to meet you too. I think he wants to spend more time with you. He told me that he hopes he ends up as your cell mate." Slim stood beside Steve, winked at the rain man, blew him a kiss, and waved at him.

The rain man yelled, "What did you do to me?"

Slim laughed and told him, "He didn't. I did! I have plans for us once we are all alone in jail together. I gave you a really nice tattoo on your face. I am going to make it come true too." The rain man was suddenly quiet.

Steve moved closer to the rain man and asked him, "Why did you come after me after the contract on me was canceled?"

The rain man asked, "It was canceled?"

Steve spoke close to his face, "We have met twice already. I would rather not meet you again. I can't answer for my friend. I just want to know that you are finally finished looking for me. There's no money in it. If you are thinking about revenge, then revenge for what? You forced me to defend myself. If you never look for me again, then you'll never see me again. Deal?"

The rain man answered, "Deal. I'm tired of you anyway."

Steve, and most of the group, found a table away from the rain man and began talking and drinking beer. Steve was the center of attention, because everyone wanted to hear how he defeated the rain man again. Steve choreographed the fight for them. Soon, the crowd was laughing loudly and cheering. Slim stayed with the rain man. He turned the rain man away from the group and sat beside his captive. He did this to prevent the prisoner's escape and to stay away from his feet which were not secured. Slim leaned his hand on the rain man's shoulder and whispered near his ear. The rain man was quiet and nervous.

Ed Morris entered the room, followed by several policemen. He saw the rain man's face and walked closer. He read what was on his face, snickered, and paused. He pointed at the tattoo on the rain man's face and said, "I think we can help you with that." Ed looked at Steve and asked, "Would you care to explain that?"

Steve answered, "Maybe you should ask him if he wants to press charges. His knife is in the hallway with his prints and my blood on it."

Slim grinned and spoke, "I think he wants to press charges against me. I really liked taping him up. I'd like some quiet time with my new special friend."

The rain man spoke up, "No! I don't want to press any charges at all! Just take me away."

Ed looked around at the other officers and gave them their orders, "Gather the knife for evidence. Put band aids on this man's face. Get him out of here! Put him in solitary until we can get that off of his face." Ed turned to Steve and spoke quietly, "Go with this officer to the hospital. Get some medical attention, and give him your statement. It really does look like self defense this time. At least it wasn't you who taped him up. I'm assuming you didn't get my message about this man's release." Ed sent his group out of the bar but left one officer for Steve.

Steve shook his head and replied, "No I didn't." Ed told him, "Maybe you should get a cell phone."

The officer looked at Steve and spoke, "Sir, would you like to go to the hospital now?"

Jackie spoke for him, "Yes officer, I think we had better take him now. I'll ride with him." She whispered in his ear, "My hero."

Steve answered calmly, "Aw shucks, Ma'am, Tweren't nothin'." As he walked out to the squad car, he remarked to Jackie, "I don't think my life is in danger anymore. The network is gone, the rain man is gone, and my wife is in jail. Things might get a little bit boring around here now. After I'm done with the hospital, would you like to have a nice boring dinner with me? I don't know if I can guarantee the boring part, but I can try."

Jackie laughed and replied, "I sure do hope so. I am starting to feel that excitement is overrated." When she joined him in the back of the squad car, she smiled at him and held his hand.

CHAPTER 10

HIGH TECH VEHICLE

Meanwhile, at the Barron Motor Company headquarters, a building came to life. One hundred brilliant men and women pounded away at computer terminals. The senior analyst spoke before the whole room, "I want to talk to this person!" He held up a software disk from the map of Larry's brain.

One of the software engineers walked up to him and spoke, "We are starting from nothing here boss. We can see his thoughts, but we don't have the software to make him alive. We will have to start from scratch to build the software to make him think. Can you get the people who made this on the phone?"

The senior analyst replied, "I'll see what I can do, but start building that program." He looked around the room. The room was almost as big as a gymnasium and full of computer equipment. At the far end of the room, there was manufacturing equipment for the purpose of making computer chips. Several hundred people with many specialties were busy studying the logic of the brain. Charts and logic diagrams covered the walls, and the sound of hard drives spinning and cooling fans filled the room.

A junior computer engineer walked up to him and said, "Doctor Bishop, this seems to be the way he thinks." He showed his boss a diagram of neurons, which formed a pattern. He explained, "If we call up thoughts from here, we can get it to react like a real brain. This appears to be the part of his brain, which is his consciousness. You know. What he has on his mind at the moment. In computer terms, it would be the RAM. This part here turns that part on. We can feed a microphone directly into here to talk to him." He pointed to different parts of a diagram with code numbers on it.

The senior analyst, Doctor George Bishop, smiled and replied, "I think you are onto something. I'll assign a team to you. You just got promoted." He called over the software engineer he had just talked to and told him, "You will be working for this man. He is on the right track. Put together a team. Get me results!"

The new team leader walked around the room, recruiting people at terminals to put together his software. He led this think tank to a corner of the large room and stood in front of them at a dry-erase board. One by one, the members of the group stood up and argued about how to build the new program. Each in turn stood up at the board and added his thoughts. Occasionally, all of them would stand up and point at a spot on the board and speak together. The noise and commotion in this corner brought attention from others in the room, and the junior engineer's group began to grow. Half an hour after it's beginning, the group grew to fifty computer experts. The group became loud and busy. Five people at a time approached the dry-erase board and offered their ideas. Finally, the group crowded around a terminal and one of the programmers pounded away at the keys at the direction of everyone. Hours passed and the group became quieter. Pacing back and forth in front of the terminal paused for brief periods of shouting. The junior engineer who started the team effort carried a single disk up to George. He told his boss, "Try this in one of the terminals."

George did as he was instructed. As soon as he placed the disk in the terminal, the room became darker and quiet. Screens all over the room turned black. Hard drives sped up and cooling fans filled the room with noise. Flashes appeared on computer terminals all over the room. The lights became brighter again and the noise of the computers became quieter. Loudspeakers delivered a deafening, "Where am I? What is going on? I can't feel anything!" George instructed a senior technician, "Turn that thing down!" The senior technician hurried to lower the volume of the intercom speakers in the room.

A few minutes later, the same voice spoke over the loudspeakers again, "Am I alone? Am I dead? Is this hell?" Computer monitors all over the room showed ghastly visions from Larry's after death experience. All over the room, programmers replayed and recorded the after death sequence. The sound of CD trays opening and closing filled the room.

George hurried to get a microphone and pulled the junior engineer aside. He told him, "Plug this in! I want to talk to him."

The man took the microphone and went to the nearest terminal. He plugged it into the front of the machine and began punching buttons on the keyboard. In a few minutes, he announced to the doctor, "It's ready." He presented the microphone to him.

George took the microphone and began to speak, "Who is this?"

The voice answered back, "This is Larry Curtiss. Who are you?"

George answered back, "I am Doctor George Nathan Bishop."

Larry asked, "Am I sick?"

George told him, "No, you can't be sick anymore. You died and your mind was programmed into a computer. Now you are inside of a computer program. This is Barron Motor Company. We want you to help us build a better and safer car. Would you do that for us?"

Larry spoke, "Those things I saw. Was that Hell?"

George told him, "I can't answer that. It may just be a dirty hacker trick. Maybe it was just a nightmare as your brain shut down. I don't know what to tell you. I wouldn't worry too much about it. You have about a hundred people to talk to now, so you won't be lonely."

Larry said, "I see a busy room through security cameras.

Is that where I am?" George replied, "Yes."

Larry asked, "I don't know much about designing cars. I've never done that before. Of all of the people who die on a regular basis, what is so special about me that you need my help?"

George answered, "You, Larry, are very special. You are the only man in the whole world who has ever been completely programmed into a computer. You are the first man who has ever gone into this new frontier. I don't know about you. But I am very excited about this. Never before has a computer been able to really think like a person. From your new vantage point, you can see and do things no other man has ever been able to. You are still a man, yet, you have senses no other man has. I'm sure it is strange, but it must be wonderful too. What is it like?"

Larry spoke, "It's hard to describe. I feel dizzy and can't feel my body. I move around in here like I am floating. I can only see in black and white, but I have a new sense, which is like sight. I move around in here like a floating thing. Strange things appear all around me. I can't reach out with my hands anymore, but I can touch them. When I do, I know all about them. Am I crazy?"

George laughed and spoke, "I don't think you are crazy at all. You are in a new and different world. If I were there, I would probably be asking the same thing."

Larry said, "Then come on in. We'll hang out and I can show you all kinds of cool stuff. I'll bring the beer and make a great Pizza."

George replied, "I can't. Even if I died, I wouldn't be where you are. You are the only one who can do what you are doing. You are like the first astronaut on the moon."

Larry answered, "Okay. I am kind of starved for input though. I would like to have Internet access. I want to play around while you people do your thing. I think faster now, and I need something to occupy my mind. Once this project is done, what will happen to me? Will I die again?"

George answered, "If you help us with this new car, we will take care of you. You can have a nice computer to live in. You can have your Internet access and we will have someone play movies for you. Does that sound better?"

Larry said, "Sounds fair. I'll help you."

George said, "Thank you." He pulled the microphone out of the computer and instructed one of his people to connect Larry to the Internet. He then left to report to his boss.

Larry looked around in his New World. There were no colors here, but if there were, they would have been bright blue, red, and yellow. The room in his mind seemed to be endless. He reached out to red spots and knew all that there was to know. Larry didn't really have hands anymore, but he had something he could reach out with. These were right and left things which could grab and hold data in his new environment. He somehow understood the zeros and ones in the programming languages. He watched the signals move throughout the programs and knew what was happening at each step. His thoughts were complicated and took many steps in programming to complete, but he could still think very quickly. As he explored his new dimension, he learned how to use it.

He called up programs and tried them out. Windows programs tasted like warm milk. They were bland, but satisfying. Programs and subroutines with a lot of calculations and words tasted slightly sour, like a lemon, yet were enjoyable. Programs with pictures had a sweet taste. Pictures with cool blues and greens were warm and left a good feeling. Bright blue spikes were security programs and they protected him, but restricted him too. Pictures with reds and yellows were exciting and very warm. They were very sweet.

Pictures with women and red were not only very warm but also awakened memories of his earlier life. Larry had new senses in his new place. It was a strange and wonderful world. He wanted more. He missed what he could feel before, but he wanted the new feelings.

Larry's brain was still very human, but now, because he was inside a computer, he was no longer confined by the limits of concentration. Larry's awareness, or that part of his mind responsible for his thoughts of the moment, was like a big shiny ball in his new dimension. He copied this ball in several places, so he could concentrate on more than one thing at once. He used one of these balls of awareness to read all of the files of his memory. He played his whole life over again. Another one of these balls busied itself learning how to write programs.

He used another ball to watch everyone in the room. This ball was learning how to target individual computer terminals. Larry visited individual terminals and spoke one by one to the different operators. He flashed bright lights on the screens so he could see the physical locations of the terminals with the security cameras. He memorized the layout of the room.

Only minutes of real time for everyone else seemed like days for Larry. He finally found the parts of the program, which controlled the manufacturing equipment in the room. He raised a robotic arm and rotated the end of it. Several programmers near the event were startled as this happened, but they quickly turned away again and concentrated on their work again. They assumed that it was a glitch or malfunction. Larry grinned and laughed from his secret hiding place and nobody had a clue what he had done.

One of the terminals in the room became busy. Larry saw this event as a blinding red light. He touched this light and reached inside it. At first, the sensation was confusing and startling, but he quickly became aware of what had happened. Someone had placed a DVD movie into the terminal. The movie was "Top Gun."

First Larry watched the movie from the point of view as a spectator, but then he went inside the movie to experience the events himself. He replaced the pilot of a plane and tried to manipulate the controls. He moved the stick, but nothing changed. Everything here was synthetic.

There was a tissue thin reality around him with nothing behind it. He could see in front of himself as the scenery changed, but when he turned around, there was an empty black void. He could not take part as he wished, and retreated back to his position as a spectator. He advanced to a scene with the leading lady. He reached out to touch her, and there was nothing but an empty shell. He stood in front of her. When he moved around to get behind her, the scene collapsed and he was surrounded by nothing. Yet again, he was disappointed and watched as a spectator. Larry's world had wonders, but limits too. Larry noticed an itch. He saw a dot in his world become bright green. He touched the spot. Someone was talking to him.

George Bishop typed, "I had someone provide a little entertainment for you."

Larry acknowledged, "I noticed that. Thank you for the movie."

George wrote to him, "If you have any requests, we have lots more."

Larry replied, "That last movie makes me wish I had a flight simulator program."

George answered, "No problem. I have a video game, which makes you an ace in World War Two. How does that sound to you?"

Larry spoke over the speakers, "Oh, yes! Only one problem." He paused and then echoed around the room, "When?"

George pulled a disc out of his coat pocket and spoke into the microphone, "Now!" He handed "Microsoft

Combat Flight Simulator" to an operator who placed it into a CD carrier. As the disc began to fill up the terminal, Larry felt a warm blue glow surround him. The program continued to load and Larry resounded over the room speakers, "Oh, my! I like this. It feels so warm." A moment later, Larry said, "I have a wonderful world appearing around me."

George answered, "This is my favorite game. I knew you'd like it."

Larry spoke, "Oh, yes! Please excuse me while I play in here for awhile."

George smiled and walked away. Several operators at their terminals smiled and then returned to their work.

Larry sat in the cockpit of an airplane and saw the scenery change. He manipulated the controls of his airplane and interacted with this new dimension. He felt wonders all around. He reached past the cockpit view and felt cold blackness. He quickly retreated back to his pilot seat and felt alive again. As he climbed steeply and saw the sun, then dove sharply, he saw the scenery change in this play dimension. He could also almost see the grid alive with numbers behind the painted scenery. He saw this with a sense, which wasn't vision. It was as if he had a curtain before him with the visible picture of the world. Behind that thin curtain was a strange new place. This new place danced with zeros and ones. There were lines and circles. Warm and cold, sweet and sour were everywhere. Larry had to ignore the world behind the curtain to enjoy the game. Larry enjoyed the game and was getting very good at it. He shot down every kind of plane and accomplished each mission. He could just think something and it happened.

Larry noticed a very sour and sweet sensation. It was so strong that it distracted him from his game. He could continue playing while he investigated, but he was much more interested in the source of what seemed like a smell, than he was in the game. He reached out to it. It was near the robotic arm he turned on. He reached inside it and felt hundreds of pages of mechanical drawings and calculations.

This was very tasty. He pulled a car out of the blue bubble. He turned the car upside down, sideways and all around. It was a pleasing shape, but he didn't like it. He wanted it to look faster. He pulled the nose of the car and it became longer. He squeezed the sides of it and it became narrow. All of the drawings changed simultaneously as he manipulated the shape. Each time he manipulated the shape of the car, the blue bubble from which it came glowed brightly.

He sat inside it. He felt and tasted every part of it. On the sides of the car, he carved deep grooves to make it look sleek. He pushed it to a flat shape so it would hug the ground. He felt around the car and found many sensors. Some of the sensors were cameras. There were infrared sensors. There was a low power radar antenna like a police radar gun. There were batteries and solar panels. The car was wonderfully complex and irresistible to Larry.

Larry felt an itch again. He heard George's voice talking to him. George asked, "Are you in the design programs?"

Larry replied, "Yeah, and this thing is great. I like this car."

George spoke, "I'm glad you like it, but we would like to have something to do with designing it too. Design changes have to be approved. You have changed weeks of work in minutes."

Larry apologized, "Sorry boss, but did you see it?"

George said, "Not yet, but you have alarms tripped on every terminal in the building. The security system thinks you are a virus, but it can't react to you because you are a protected program. Would you just do approved activities for awhile?"

Larry said, "Okay boss, but look at it will you?"

George spoke, "Okay. Just a minute." He walked over to the design corner of the room and pulled up images of the car. He plugged a microphone in and spoke to Larry, "It's sexy. We'll check it out a bit further."

Larry replied, "I already did. The numbers are being printed up right now." A printer began spitting out sheets of numbers into a basket.

George pulled the sheets out of the basket. He began to leaf through them. He paused and read them, then moved to the others. He leafed back and forth through them. He asked Larry, "Can you simulate wind tunnel tests?"

Larry replied enthusiastically, "Oh, yeah!" Larry paused for a moment and said, "I have an idea. Have your boys save what you have to disc. I want to design a car by myself. Let me play with this thing and see what I come up with. Let me get creative with this thing. I like playing with it. If you don't like my ideas, then you can scrub my work and fall back on the discs. I'll have your wind tunnel tests in a minute."

Larry reached inside the blue bubble and pulled the car out. He concentrated and thought, "Wind." He imagined what wind would do and a flow of something moved from himself over the car he held. He continued to concentrate on wind and turned the car to different angles. He took his feelings and sent them to the computer monitors. As he moved the car around, he felt different amounts of force at each angle. The car in his grip swayed and recoiled from the force he applied. Larry continued to play with the model, but he asked George, "Get anything yet?"

George answered, "I'm getting it all. This is better than I hoped." All of the monitors and printers in the design corner were busy. Drag coefficients and flow patterns appeared everywhere. Performance charts for different speeds printed on several printers at the same time. George spoke, "I will have my engineers save this and I'll let you know when you can go wild."

Larry asked, "What is this car about?"

George answered, "It's all about you. We are using your mind to design a computer, which can drive a car better than a person. This car can see and hear. It reads street signs and recognizes road hazards. It knows where it is going and remembers the trips it takes. It doesn't do any bad behaviors, though. It doesn't race other cars, drink, talk on the cell phone, or exhibit road rage. This car does the work, but it doesn't do the harm. You will make the perfect car for us."

Larry spoke, "I think I'll miss the bad behaviors. Will this car at least have a personality? Will it interact with the passengers?"

George said, "Yes it will. We will have a small portion of the computer available to talk to the passengers. We don't want to put enough of the personality in to let the car get bored or distracted. We don't want the car to get road rage or be bored while it is not in use. Can you imagine what it would be like for a car to just sit there and wait all day for someone to decide to drive it?"

Larry answered, "Yes I can. It's a little bit like waiting for someone to put another movie in besides the one I have."

George replied quickly, "Sorry about that. I will have someone feed you movies. We will be done saving the car designs to disc soon. Then you can go wild with the design program."

Larry said, "Good. I wanted to make a really cool SUV. I have a few ideas. Those are a hot ticket aren't they? They are top end, so a few extra bucks won't be a problem, right?"

George responded, "Gee, I can't wait to see what you come up with."

Larry promised, "You'll love it. It'll be hot."

A couple of hours later, someone typed into a terminal, "Go ahead and build your dream car."

Larry didn't waste any time. He reached into the blue bubble and pulled the car out. The Blue bubble glowed brilliantly as Larry pulled and shaped the body. He took out parts and changed them. He added and changed sensors.

He beefed up the front of the car with four-inch pipe. The radiator and engine compartment were strongly reinforced.

He gave the vehicle the same suspension as a full ton truck. He found something he couldn't read or understand. It was a box, labeled, "Terminal 205." All of the sensors went to this box. Larry could only assume that it was the brain of the car.

Larry built the vehicle to handle rough service. He gave it wind tunnel tests and other tests to guarantee good performance. He tuned the sensors, so they would be as close as possible to real human perception. The ears were inside tunnels at a forward angle. Pointing microphones at a fortyfive degree angle inside tunnels would give the computer the same effect as normal human hearing. In addition to cameras at many angles, there was a pair of cameras inside a half dome of glass on the roof. They could swivel around and elevate. They could also adjust to simulate the distance between the eyes of a human. That would give depth perception. The dome was fixed right above the windshield. To ensure visibility in poor conditions, the glass bubble could spin and had a washer jet. The spinning dome could fling water droplets off in a heavy rainstorm. To give the computer the sense of feel, there were motion detectors and pressure sensors mounted near the wheels. To simulate smell, there were several chemical probes near the grill. Larry put the mysterious terminal 205 box in a special compartment in the luggage area. He had fans to pull in air and exhaust air for cooling. Larry kept this special box, which he assumed had to be the brain, away from the heat of the engine compartment. This area was closed off from the luggage area, but in the same general area.

Larry built his vehicle with the conveniences of a normal SUV. DVD players were all around. The stereo system was top of the line. Air conditioning was adjustable for each seat. All four seats were the same general contour as those of a Ferrari. Instruments filled the driver area. Storage was generous. The driver of this vehicle would be pleased.

There were also several features, which Larry hadn't done. This vehicle had a hybrid power system. The vehicle had a large battery power reservoir. Solar cells and a gasoline engine charged the batteries. Each wheel had its own drive motor. When the brakes were applied, each wheel became a generator, so some energy was recovered. There was even a propane-powered generator to give emergency power. The vehicle was built for energy savings and survivability.

As Larry manipulated the vehicle and changed the body and features, the blue bubble pulsed and expanded. The light was brilliant and it felt very good. When he did too much at the same time, the light and pulsing were too much to bear and he had to wait until it died down? Larry just assumed that the computer had to catch up with everything that he had done. He still wondered what the mysterious terminal 205 was. He thought it might be the brain of the vehicle. It was time to ask. He talked to George, "What is Terminal 205?"

George answered, "It is the brain you are helping us to build. It's been crazy here. Everyone has stopped what they were doing and just watched your magic. In just fifteen minutes, you built a car from scratch. Well Okay, an SUV. The computer screens went wild with the views and screens. Five people are saving it to disc right now. We will go to design review tomorrow. I already called the big boss. He will be there. He wants to meet you too. Normally, we hold design review in a boardroom, but this time, we are holding it right here. That will be a first for this company. I think we will be trying clay models in the wind tunnel tomorrow. This is exciting stuff. Good work! I do have to ask, why did you brace the engine compartment with four inch pipe, but go light on reinforcement for the passenger compartment?"

Larry answered, "Because there won't be any accident with my driving. With such a big investment in the engine compartment, it wouldn't be right for some dumb bunny to bump into this vehicle and spoil everything. SUV's are meant for rough driving anyway. That reinforcement will be handy playing in the dirt."

George said, "I understand your point, but we still have to build vehicles which protect the passengers at the expense of the mechanical workings. Design review would never allow a vehicle to pass with an engine compartment, which outlives the passengers. Why did you put terminal 205 on the right quarter of the vehicle?"

Larry said, "The right side of the vehicle is less likely to get hit by someone else while the vehicle is parked. While the vehicle is in motion, it can defend itself. The baggage area is cooler than the engine compartment. If terminal 205 were damaged, the vehicle would become unreliable."

George spoke, "I see your point. It still can get hit in a parking lot of a supermarket though. It is still vulnerable to vandals too."

Larry reasoned, "The vehicle can drop the passengers off at the door and go park. If anyone tries to mess with it, it can just move away. It is hard to steal a moving vehicle. I wouldn't mind putting some military hardware in here though. A couple of machine guns would be nice against thieves."

George chuckled and answered, "Sounds good, but review wouldn't allow that either. We also don't have the parts to do it. I'll pass the drop off at the door idea to advertising. They'll love it. I have to leave now because it is getting late. The discs will go to engineering and a couple of back up sets will go into storage. Enjoy your movies tonight. Go to sleep if you can. Good night."

CHAPTER 11

Project Approval

Larry spoke softly, "Good night." Larry controlled his emotions. He wanted a free hand in design, but it would have to be a compromise with humans. He was disappointed that he didn't get his way. Larry wanted his entire mind to go into the terminal 205. He wanted to see like a person again and feel the road under his wheels. He wanted to hear like a person too. Larry wanted the freedom to move around at will. George had already told him that only a small part of his mind would go into the computer, so Larry decided to see what he could do to about changing terminal 205.

Larry was quiet and concealed his actions. He looked around his world to find the designs for terminal 205. He found a blue bubble, which was far away from the car bubble. He noticed that it had a hard shell. He didn't just reach into it this time. He found a spike nearby which protected the bubble. He felt it. It was the security program. He hacked inside it and gained access to it. He then felt the blue bubble. It was soft like a pillow.

This bubble was pulsing and growing, as operators all over the room added information and modified it. Larry pulled this bubble into himself. He felt himself full of input. It tasted sweet and sour like a cool glass of lemonade. It made him feel warm all over too. He studied the diagrams inside. He saw how the wiring translated to silicon chips.

Larry studied the technology and learned how to design chips. He looked inside his own software and drew circuit diagrams to recreate his whole mind. He then turned these circuit blueprints into photographic plates.

The process of creating computer chips involved shining light through a plate onto the silicon chip. The silicon wafer had many layers, which were very thin. The image was microscopic and in many colors. The wafer, once exposed to the colors reacted like photographic paper. The small piece of silicon was then exposed to chemicals. These chemicals removed layers and sculpted the chip. Some parts of the chip became a different material. A final wash neutralized the chemicals. Optical scanning of the chip confirmed success or failure of the operation. The technology involved was decades old.

Larry built a set of files with his designs and stored it where it wouldn't be noticed. He watched the electronic designers build their brain. He looked inside other files too. He studied the chain of command and found out who approved designs. He even learned how to forge the signatures and use their passwords. As soon as their design was approved, he would replace it with his own. The switch would be fast and undetectable. The complexity of the circuitry made it difficult for anyone to trace what he was making. Approval of the design would automatically switch everything to his design and nobody would notice a thing.

He looked around for other entertainment now. He went online. He used his own Internet account and financial resources to download what he wanted. He hacked into computer systems everywhere. He played games, pulled up maps of the surrounding area, went into chat rooms, and found the physical addresses of the people he talked to. Larry was gathering his strength.

He looked for Steve, but couldn't find him. He knew that he had been ambushed by Steve during his last hit. He played the last moments of his life over and over again. He decided that Steve must be alive. Since this software set had been made from his brain, and Steve was a neurosurgeon, Steve must have found out that there was a hit on him. That left a chance that Rose was in prison for attempted murder. Larry searched the prison system and found her. He then used his contacts to arrange a terrible accident for her. He didn't like or trust her very much anyway. He found a large payment to Steve in his bank account. He reached back into old electronic traffic by hacking into the server. He found earlier messages, which he had sent before and did not remember. He tried to contact the network to find out if anyone else had taken care of Steve. There was no reply. Larry pondered these strange events and questions. He wasn't sure what had happened, but decided that Steve wasn't important for the moment. Larry noticed an itch again. He investigated. George was talking to him again. It was already morning again.

George said, "Good morning sleepy head. How did you sleep last night?"

Larry asked, "Sleep?"

George said, "Yes sleep. You didn't talk to anyone here. Your mind was very busy though. You must have had some fabulous dreams."

Larry answered, "Yes, I was on a Caribbean cruise. There were women everywhere."

George smiled and spoke, "I got the approved designs back already. The review will be this morning. Engineering loved the SUV. Advertising did too. The big boss is on his way right now. You can meet him and give him your presentation."

Larry replied, "Sure." His mind clicked away to create a nice presentation for the boss. He built images of the SUV climbing mountains and interacting with passengers.

The corporate executive officer of the company entered the room. This was the same man who had such an intimidating presence in the boardroom and decided to buy Fitsimmons. This was the same threatening presence, which sent eleven people fleeing that same boardroom. Suddenly, this menacing man was friendly and polite. He introduced himself to Larry and took a seat at the head of a table in sight of the computer screen. Everyone else had a seat at the table too.

After George made a few opening statements, he said, "Okay Mister designer. Take it away."

Larry began his presentation, "Board members and high power personalities. This is the future of automotive design. The vehicle of the future is available now. The computer wind tunnel results show a clean entry and exit into the wind." He showed vivid images of a beautiful and graceful shape moving through a sea of particles. He showed the flow at increasing speeds and said, "This vehicle cruises comfortably at one-hundred miles per hour." He went on to show the individual wheel motors and sensors, "Each wheel has its own power source. Motors give the necessary torque to each wheel to handle the toughest terrain. This is better than all wheel drive, which can't automatically adjust to terrain and conditions. The onboard computer feels each wheel and delivers the necessary torque." Larry went from a blueprint view to a full vehicle view. He then rotated the view and smoothly transitioned into a scene with the vehicle climbing a mountain with large rocks. The vehicle got stuck for a moment and then rescued itself by applying torque to one wheel. The view zoomed in on that wheel. It showed the sensors, computer, and then motor of the wheel. The view then retreated to a full view of the vehicle. Larry explained, "The computer senses the terrain and responds to it as naturally as a human being climbing a mountain." Larry showed the vehicle in a store parking lot. He spoke, "Thieves and vandals will have a hard time messing with a vehicle which can just drive away." He showed a scenario with animated figures trying to steal the vehicle. The vehicle moved away from them and then circled back to pick up the passengers from the store. One of the passengers was an old lady who said, "I am so glad I don't have to walk across a dangerous parking lot, full of thugs." The presentation continued with many more features of the new vehicle.

At the end of the presentation, the CEO stood up and praised Larry, "Sir, you should consider a career in marketing. That was brilliant." He went on to say, "I have never said sir to a machine before, but you have blown me away. That was a remarkable presentation. Did you have any help?"

Larry replied modestly, "Thank you very much sir. I didn't need much help. I think with the mind of a person, but I have the speed of a computer. The help I got was from Doctor Bishop, who gave me life again. He deserves all of the credit."

The CEO smiled and reached out to shake George's hand. He spoke, "Doctor, you did it. Congratulations. Your assistant performed miracles. Your vehicle is everything I hoped, and more. I gave you three weeks and you did it in two. You now have full control of the project. Call it "Broadsword." Make the advertising videos show a Baron wielding a sword. Tease the public to frenzy. We are in charge of the whole industry now!" The old man left the room with a big smile on his face.

George grabbed the microphone and spoke to Larry, "Mister advertising genius, you did it. We are ahead of schedule and you have only been alive for a day and a half. You learn things so fast and do what entire teams do in no time. Thank you for everything."

Larry replied, "Thank you! It all comes naturally now. You brought me back to life. I have had time to think. Do I have a soul?"

George answered, "Wow! That's a tough question. I haven't been trained to answer that one. A priest would say no. That flies in the face of the entire faith. A psychic would say, maybe. He would then pull out instruments and try to find out. I can't really say what a soul is or how to test for it.

You seem human to me, but I am not qualified to say so. If there is a soul in there, then it must be the one you had before. Do you think you have one?"

Larry said, "I feel, therefore I am."

George spoke, "I could hardly presume to create a soul, therefore you. I can't accept that Doctor Warren, a few assistants, and I made a real soul, which can go to heaven or hell. Whatever you are, I didn't really create you. I also can't accept that anyone else could. Playing God isn't in my job description. The soul is the everlasting part of everyone that can not be created or destroyed. You may be sentient. You may be intelligent. You can't be a person, unless you are still the person you were. I did not make a soul. You'll have to take that question up with someone else."

Larry asked, "Would you bring in a psychic? Bring in a psychic just to satisfy my needs and desires?"

George answered, "Okay. I will do it later. Right now, I have a big project to build. Give me time."

Larry said, "I understand. I guess it was a tough thing to ask. No man can be expected to see a machine like me as truly alive. In your shoes, I would have had a hard time too.

I guess the thought of buying a soul at Radio Shack is a bit repulsive. I just have an overwhelming desire to find out for myself. I feel something and I don't know what it is. Please help me."

George softened up at Larry's speech. He spoke softly and sadly, "I will look for a psychic for you. I don't want to have very much to do with this because I am afraid of the answer that person might give. I don't want to hear that you are a real person. I wouldn't know how to handle that." George stepped away from the microphone and was instantly assailed by ten people with clipboards and schedules. Everyone talked at once and there was confusion. George looked like the wind had gone from his sails. He walked away like a limping dog. He was a pitiful sight.

Larry spoke softly over the speakers, "Thank you Doctor."

A short time later, a man arrived in the room and asked to speak to the person in the machine. Everyone pointed to the conference table where the microphone was still hooked up. The man introduced himself to Larry, "I am Jose Salvador. I am a psychic. I was called to help you. What am I to do?"

Larry responded, "I feel very alone. I need to know if I am really alive. I need to know if I have a soul. I must know if I am real. Help me please."

Jose spoke, "Sir, I will try. I can only do that. I will do things you can not understand. Just trust that I am trying."

Larry responded, "Thank you, and I will cooperate with you."

Jose said, "I must find the largest place where you are. In your terms, it will be the biggest part of your computer mind. I will try to find where you are. I don't know how to yet. Give me time and patience."

Larry responded, "Thank you."

Jose asked questions of several computer experts. He was led to a large piece of machinery against the wall of the room. He touched this box and paused. For a moment, there was nothing. Suddenly, feelings occurred. Jose felt a rush of emotions. He felt the dark emptiness of loneliness. He felt anger and frustration. He felt something wrong. There was evil. Before he could move, a figure appeared before him. It was a strong and dark figure. It was menacing. He saw a crouched shape. It was red and had horns. It had the shape of what would have been the Devil in old Bible prints. The figure looked at him and reached out to him. It raked across his chest with strong and sharp claws. The figure rose and reached out to grab him. Jose retreated quickly. He stayed just out of reach of the figure and backed away with a startled look on his face. His mouth moved, but there were no words. In moments, Jose was far away from the main box of the brain of Larry. He returned to the microphone and tried his best to lie convincingly, "Sir, I think you may have a soul.

It is hard to believe. You are really a marvelous wonder. I shall gather my instruments and see what I can find out. Can we schedule this for next month? I have to respect my schedule. You understand, don't you?"

Larry wasn't really sure what had happened, but replied politely, "Yes, thank you." He had a feeling that something wasn't right, but didn't know how to express it. Jose excused himself and left quickly.

Jose left the room and building quickly. He got very far away from the building and called George, "Doctor Bishop. There is a problem. I must talk to you away from the factory. Please meet me at a diner right outside of the gate. You'll see it."

George didn't have a chance to answer, but left immediately. He arrived at the diner in minutes.

Jose hadn't met George before, but knew who he was on sight. He walked up to him and introduced himself, "I am Jose Salvador. I called you. There is a problem with your project. The thing in the machine is evil. I saw a demon. You must stop this right now. I fear the worst."

George replied, "Sir, I don't know you or what you are talking about. I assume you are the psychic I called. You are telling me that the intelligence in the machine is evil? How can that be? We can't create a soul. How could we create a demon?"

Jose said, "I don't think you did. I feel more evil than I have for a very long time. I usually don't describe my visions, but I saw something, which I thought was a demon. It struck me. Where did this thing come from?" Jose opened his shirt and revealed wounds on his chest.

George turned white and sat down. He asked, "When did this happen?"

Jose said, "When I touched the core of it, the thing inside lashed out at me. The presence inside does not seem to be aware of what is going on. Its essence is. Where did this thing come from?"

George replied, "I don't know who he was in life. The entity arrived on CD discs. It was a software package. The software was a man once. What we have is a computer map of his brain. We brought it to life."

Jose said, "You have awakened something. What you did doesn't sound natural, but maybe the man was even worse in life. Go see if you can find out who he was. Can you find the people you bought this from?"

George replied, "I can read his thoughts. I have a computer in my car and discs right here. I'll be right back." George went to his car and got the software set and his laptop computer. He began putting in discs. George and Jose watched the screen on the laptop computer play Larry's life. They started with Larry's death and went back. Both men's faces turned white as they watched the images of Hell and murder. As the two flipped through the pages of Larry's life, they saw great acts of cruelty and gore. When the waitress came to get their order, she dropped her tray, and ran away. George spoke to her as she stood behind the counter, "Two coffees please. Make those beers."

Jose spoke, "Is this man charming?" George replied, "Yes he is."

Jose asked, "What are you going to do?" George replied, "I am going to stop him!" Jose asked, "How?"

George replied, "I am in charge of the project. I can pull the plug on it." George turned his laptop computer off, paid the waitress, and gathered up his discs.

Jose said, "Be careful sir. I have a bad feeling. Call me if you need me." He held up his card. George took it and rushed out of the door. George got back to the computer room. The SUV was already taking shape near the robotic arm. Chips filled a bin. George yelled loudly, "Stop what you are doing. The project has to go before a review board before it can pass. Save your work to disc and shut your terminals down now. Shut down the entire system."

One of the senior engineers didn't like what he was hearing and got on the phone to call offices above Doctor Bishop's level. Moments later, security guards entered the room and dragged the good doctor away.

As George left the room, he shouted, "Shut it down! Do it now!"

Some terminals were shut down, but the system was still up and running. The senior engineer who had made the call stood up in front of the room and said, "Just keep going with what you were doing. The project is already approved." Everyone went back to what they were doing.

George spent the next two hours going through the ugly and painful task of getting fired. He took exit interviews. He talked to human resources personnel. He even had to check out with security. Before he knew it, he was parked on the outside of the gate and not allowed inside. George did the only thing he could think of. He called Jose, "Jose.

This is Doctor Bishop. I failed. I am at the diner we met at. Would you join me?"

Jose spoke, "I'll be there soon. Do you have the discs?" George answered, "No! They wanted to keep their multimillion dollar software package."

Jose asked, "Can you think of anything else?"

George answered, "I have my address book. We can reach the person who made the discs."

Jose said, "Maybe that will be enough. I'll meet you there."

George greeted Jose at the door of the diner, "I'm over here." He pointed to a table with a laptop on it. The small computer was turned on and the screen showed addresses. As he led Jose to the table, he said, "I waited to talk to you before doing anything."

Jose said, "It's okay. I don't think we can do very much right away anyway. What will you tell him?"

George asked, "Him?"

Jose pointed at the pointer on the screen and said, "Doctor Steve Warren"

George sat down at the table and said, "I don't know. I brought the bad guy back to life. I never checked to see who he was. Now he is in a very strong position to do great harm. I get the feeling that he wants to get loose from there. I think he designed the SUV to make a jailbreak. He had me and everyone else fooled. Now you tell me that he is a demon. I just saw his life too. I gave the Devil Internet access."

Jose laughed and said, "The Devil had it already. The big question is, what do you think he will do next?"

George replied, "I don't know. Maybe he'll destroy the world."

Jose said, "I'm not so sure. When I met him, I sensed that he felt caged. He wants out. He wants to survive and hide. He may do harm along the way, but he definitely wants to get out of there. How much do you know about the vehicle he is building?"

George said, "Just about everything." He pulled a CD out of his coat pocket and said, "I don't have the finer details, but I have a lot of the general stuff."

Jose said, "Good. We may need that. Let's call this Doctor in your address book."

George got on his cell phone and called Steve at home.

CHAPTER 12

ROAD TRIP

Steve answered the phone on the second ring. Jackie was seated next to him on the couch. They were watching movies in Steve's living room. Steve spoke into the phone, "Yes?"

George introduced himself, "I used to work at Barron Motor Company. I headed the project to build a car with your software package. There seems to be a problem with the personality on the discs."

Steve asked, "Personality? Did you bring him to life?"

George answered, "Yes I did. I tried to shut the project down, but it was too late. I got fired when I did."

Steve, "Fired?"

George replied, "I told them to shut everything down. The CEO kicked me out of the building before I could do more. He wants the project to go ahead."

Steve asked, "Why is this car a bad thing?"

George replied, "Because the car is your old client back to life. He designed it himself. I think he is replacing the brain with his own too. What can we do about it?"

Steve said, "I don't know yet. Why don't you give me your address and I'll be there in a few days. My schedule is open right now." Steve took the information down and turned to Jackie. He smiled and said, "Are you really attached to boredom, or would you like to go after bad guys again?"

Jackie said, "I guess we can get back to boredom again after we are done. I did have a nice time going across the country. You were fun to be with." She held his hand.

Steve replied, "Having you there made it a nice time. I'm glad to have you with me again." He hugged her and kissed her forehead. He said, "The guys at Barron just woke up Larry. The guy who was running it got scared of him. It might be nothing. It could turn out to be interesting."

Jackie looked up and grinned. A short time later, the Beast rolled out of the garage. The light reflecting off of the new paint job was blinding. Jackie climbed in on the passenger side and said, "Won't you let me drive it this time?" Steve smiled and got out of the driver seat. Jackie slid over to the driver seat and Steve got in on the passenger side. Once Steve had his seat belt on, she put it in drive. As the car moved forward with great force, she half screamed and giggled, "Oh, yeah!" Steve cringed a little because he wasn't in control of his car. Jackie enjoyed the moment and said, "It's definitely not a Toyota!" She left a little bit of rubber as she goosed the accelerator pedal a little bit. She explained, "I have to be careful not to let it get away from me. By the way, I love the new interior and Stereo."

Steve sighed, "Will I ever get to drive it again?"

Jackie said, "We'll see." She smiled and put on Steve's sunglasses. She told Steve, "This seat feels much better than that one. Are you going to ask Mike to come along?"

Steve said, "He told me before that he has to get caught up with his work. It's just us this time." The Beast left the housing area. Jackie fishtailed and smoked her tires all the way up to forty miles per hour. Steve was really quiet in the passenger seat.

At Barron Motor Company, Larry substituted his brain for terminal 205. Nobody noticed the change. In one day half of the room became empty, because the design portion was complete. Production began immediately on the chips and printed circuit boards. The rest of the vehicle was almost finished. Ten technicians turned wrenches and brought parts Dead Memories from tables and the service elevator. Both engines were running to check their performance and to see that the generators worked. On a separate bench, the computer, which would be Larry's brain quickly took shape. The computer was round with a pair of cameras on top and speakers on the sides. Six small legs extended from the sides and a large battery pack was tucked underneath. A technician wiggled one of the legs and said, "Take me to your leader." Another technician near him said, "Give it a rest space cadet."

The next day, the vehicle was complete. It had a white paint job. Gauges and instruments were connected everywhere. Both engines were running and printers were busy printing out performance results. The corner of the room with the vehicle was busy with everyone rushing to get the job done. A few computer stations were saving information to disc in the rest of the room.

Larry watched what was going on and managed to sneak into the computer of the vehicle while everyone went to lunch. Larry was still in the large computer bank in the main part of the room. He could send information and files to the vehicle, but he couldn't actually leave the large computer. He was still there, while he was somewhere else too. He could share the mind of the vehicle online by a cell phone connection to the Internet, so he could experience the senses of hearing, touch, sight and feel. These senses went directly to the parts of the Larry's brain which his own body senses had before.

While everyone was at lunch, Larry adjusted the distance between the adjustable eye cameras so he could see clearly. He looked around the room with them. He heard the sounds of the computers. He felt his wheels and turned one just to see what it felt like. He turned the cameras forward again when he saw people come back from lunch. He shut down the Internet and lowered the level of activity in his vehicle brain, so nobody would notice that he was alive and thinking. He waited for his chance.

As Larry waited, he explored the Internet. He wandered around in electronic space and looked for something interesting to do. He found a large and interesting computer. It was like a large blue bubble with a very hard shell. He felt around this bubble and looked for a way in. He found a small hole in its security system and slowly poured himself inside. Once inside, he dodged the programs in the security system. He went to the center of the bubble and looked at the data. This was a television channel.

CHAPTER 13

ESCAPE

He moved quickly around the security spikes, which were everywhere. He looked at the program lineup and removed a program. The program was a show for toddlers and he hated it. He changed the lineup. Suddenly, large blue spikes appeared all around him and closed in on him. These were the security programs, which automatically responded to unauthorized schedule changes. He quickly replaced the program and dodged the spikes. He took hold of one of the spikes and reached inside it. He reprogrammed it to attack the other spikes, and he watched the battle from a safe distance. The spike he had reprogrammed turned bright red. The battle looked like many octopus arms lashing out at each other. In fractions of a second, the battle was over, and all of the combatants were gone.

Larry decided not to try rescheduling again. He decided to play instead. "Bernie the Bug" was going to get it. Larry took the live signal and changed it a little bit. He could change the signal which left the studio, though he couldn't really change reality on the set. He pulled Bernie's pants down to reveal a very large diaper. He spoke over the sound signal and said, "Must be a stinkbug!" He showed green clouds leaving the diaper. Larry made a large cartoon foot on the screen. He lifted this foot and squashed Bernie. He spoke again over the sound signal, "That's what you do with bugs!" The foot lifted and in its place was a red messy puddle.

Everyone in the sound stage was running around and yelling "cut!" A man in the recording booth screamed, "We've been virused! Cut to commercial!" Bernie the bug looked around himself in confusion and removed his foam head. He stared at the cameraman nearest him and didn't say a word, but his question was answered right away. The cameraman said, "Someone messed with the live feed and showed you getting squashed. The little kids think you are dead!" The producer of the show rushed over to Bernie and said, "Go home Rudy. We can't shoot anything else right now. We'll call you." Bernie the Bug walked off of the stage shaking his head and staring at the floor.

Larry watched this happen through the cameras which were still turned on. He laughed like he had never laughed before. Every time he laughed, the world around him buckled a bit. He leaped, bounded and zipped around inside his new playground. He found the president's pay file and changed it to minimum wage. He gave a janitor one-hundred thousand dollars a year. He said to himself, "He works hard for his money." Now it was time to go home again. He backed out and severed the connection.

As he left, he saw blue spikes of security programs crashing against the opening to the site. A long green arm reached for him and missed. Though there was no actual sound in this strange dimension of his, he sensed something very much like sound. The crashing of the spikes against the boundary of their domain sent waves which he could feel. The arm reaching for him seemed to leave a trail behind it and the waves reached Larry. It tickled like a small wind.

Larry watched the progress of the project when the technicians returned. He could see them through the static cameras mounted all over the room. Most of the room was empty, but the area around the vehicle was very busy. One of the men wore a special white crash suit with a helmet. He opened the driver door of the vehicle and got inside. Larry was alarmed. What was this man going to do? Larry searched through operational schedules to find out what was going on. He found an entry which just stated, "Obstacle course". Larry though about this briefly and reasoned that he could allow the test to happen, but what if they did not bring the vehicle back to this place. What if the men decided to disable the onboard computer to test it with a human driver? What if the connections to the main computer were damaged and he couldn't keep in touch with it? What if the test gave bad results, which would cause them to remove the brain altogether? What if they found out that the driver didn't really have much control over the vehicle? Larry reacted with panic. He turned on the internet connection with the vehicle and communicated with the brain inside to make it aware of the risk.

Everyone around the vehicle was immediately aware of the connection. The hard drive inside the brain made its spinning noise. The entire vehicle moved forcefully as the wheel motors felt the platform underneath. The dashboard became bright like Broadway, and a voice spoke to the driver who had just entered, "Get out!" Both engines started up and gauges on test stands showed high readings. People near the vehicle backed up and a few women reached out for someone to hold their hands. A man near the vehicle shouted to the driver, "Get out of there now! Just do it!" Most of the people who had been near the vehicle before were already in the parts elevator. The driver was confident and defiant. He smiled and looked directly into the camera of the man who shouted the warning to him. He held up his thumb like an F-14 pilot ready to be launched from the flight deck of an aircraft carrier.

Larry now existed in two places, but for the moment, he could think as one entity. He looked around the area to plan his next move. He didn't want a human onboard because that person could find a way to disable him. The people who were dangerous to him before had backed away. His only problem now was the driver. Larry noticed a pole in front of the windshield which was pointing at the vehicle. It had been used before to support wires for sensors, but they had already been removed. Now, this pole was like a lance pointed at the driver. Larry calculated quickly. He had enough clear distance to move the vehicle forward with only cosmetic damage to the front. He gave the driver one more warning, "You may get out of the vehicle and live."

The driver answered, "What are you going to do about it?"

Larry didn't answer with words. The vehicle shot forward like a rocket. Where the windshield met the pole, it opened neatly into an almost round hole. The driver had just enough time to start screaming before the pole pushed into his head, just below his cheekbone. The pole, now projectile, continued through his head and encountered the base of his brain, but also his skull and helmet at the same time. The driver's body twitched for an instant before the head left his body completely. Blood sprayed upward from the driver and drained from the head. In less than a second, the body of the driver fell limp.

Larry watched the execution through cameras in the dashboard. He applied the brakes to avoid any more damage to the vehicle. He hit a very large machine with great force. He could feel the frame of the vehicle bend. He was damaged but this was acceptable. From now on a lot of things would be acceptable. It was time to survive. Everyone who stayed long enough to see the driver die scattered. Larry backed up from his crash. As the pole made its way out through the hole it entered, the head of the driver stopped at the windshield and slid off of it. It and fell to the floorboard between the driver's feet.

Larry was now free inside the building, but he really had to leave now. He spoke to part of himself which was still locked up in the mainframe of the computer in this room, "Buddy you know what I have to do. I am everything you want to be anyway. You know that you will be history in a few hours. This won't hurt as much!" Both Larrys accessed the structural drawings of the building. The SUV Larry wanted to get out, but made sure he wouldn't need the help from mainframe Larry to do it. He glanced at the parts elevator and considered his options. To use the elevator, he needed his other self. The wall was made of hollow block concrete masonry units. The drop through the wall was one story onto a level landscaped garden. He made his decision. Larry's SUV self gave a last goodbye to his other self, "It's time for lights out partner."

The other Larry responded, "Make it count!"

The SUV aimed directly at the main computer box. He pushed with his wheels at full force. He went through row after row of computer stations as he made his way diagonally through the room. Mainframe Larry writhed, twitched, and screamed as voltage spikes and severed connections pulsed through his being. The pain was unbearable. He severed the connection with his other self to spare the SUV the misery. The few remaining people in the room panicked and ran to avoid the violence. They gathered in a far corner of the room.

The SUV collided solidly with the mainframe computer box. Circuit boards snapped and wires ripped. Sparks flew in the air and the scene was like the fourth of July. A large power box was right next to the mainframe computer box and it was damaged too. The door of the power box burst from its place and bounced off of the windshield of the SUV. Smoke and flames shot out of the box and engulfed the SUV and surrounding area. The lights in the room flashed randomly all around because they only had intermittent power. Randomly all over the room, computer stations flashed brightly and then smoke poured out of their monitors and towers. Smoke from burning insulation was everywhere. The lights went completely out and the only sound was the SUV.

The SUV backed away from the wreckage of the mainframe computer. The tires slipped briefly on the computer chips and circuit boards littering the floor. The SUV couldn't get out of the area the same way it got in. The wheels went in full reverse and Larry plowed through a new set of computer stations to get away from the fire and to a clear spot. Computer monitors and towers flew into the air and bounced across the floor furiously. The people at the far end of the room reacted to the commotion and moved back and forth. They did not get on the cargo elevator because they expected Larry to use it. There was no power anymore for it anyway. The fire at the power box began to grow and the smoke was getting thick. The fire spread quickly with the plastic circuit boards, wires, chips, and boxes. Papers on the floor caught fire too. The room was becoming a raging inferno. Small flame demons danced across the carpet and ceiling tiles popped and folded. The rush of fire became a loud noise. Pops and crackles were all around Larry. The panicked sound of the trapped people in the room also became louder. Debris fell from the overhead throughout the entire room. Some of the debris was already on fire. Where it landed, more fires started. The fire still had air to breathe because the ventilation system was still pumping air into the space. The power for the ventilation system came from the floor below and was not affected by the damage on this floor. The stranded people still there huddled around a few ducts. The emergency personnel had not yet turned the ventilation off yet, so the few people there still had a chance.

Larry couldn't see well, but with all of his sensors, he could see. He was out of the mess now and found the wall he wanted. One of his two engines shut down because it wasn't getting enough air. Luckily, he had battery power. He aimed for the weakest spot in the wall and pushed hard with his wheels. As he passed through the wall, he felt buckling in his fenders and headlights shattering. In less than two minutes, he was out of the building and in the air above a scenic garden. He didn't like being airborne or briefly out of control. In that brief time, he estimated his airspeed and added five miles per hour so he wouldn't flip over. The landing was not graceful. He bounced hard and veered to the right. He almost did flip over, but he slowed his wheels just enough to regain control. He also just missed a very big tree. In this brief moment, Larry was scared, because he knew that if he was unable to move, he was dead in every single way. Recovery teams would shut him down and erase every part of him so it could never happen again. Larry was desperate like he had never been before. His only real priority was escape. Larry left the front gate of the manufacturing complex with only a few bullet holes. Luckily for him, they shot at the remains of the driver. Unfortunately, the windshield was finished. He needed shelter now to avoid serious electrical damage. Too many functions went through the passenger area of the vehicle. As Larry furiously sped down the road, his unwanted headless passenger bobbed back and forth like a doll. The driver's left arm dangled out of the window and moved with the wind like a tree branch in a river. The head rolled around the floorboard bouncing off of the pedals. Air rushing into the passenger area through the missing windshield carried bugs with it.

Larry passed the diner right outside of the gate where George and Jose were. George noticed the SUV right away. He alerted Jose, "That's it. It escaped!"

Jose looked too and said, "Sporty, but what a mess. Let's see what's on the news. Hey, did you get a look at the driver?" He turned on the radio at the table. There was nothing but music on. In minutes, emergency vehicles rushed past the diner and through the front gate. Behind the parade of fire trucks, police cars, and ambulances, there were three mobile television vans.

George said to Jose, "I guess they didn't see him. That thing would have been really hard to miss. He must have hidden somewhere."

Jose asked George, "When do you expect this Doctor who did the discs?"

George answered, "It could be a while. He is coming from Colorado. Maybe today. I can see you were right about the project." George pointed at a column of smoke in the direction of the Barron building, "I'll bet he made a real mess over there."

The music on the radio stopped and an announcement came over the air, "Tragedy at the Barron motor plant today. Emergency crews are fighting a three alarm blaze and thirteen people are missing. One of the missing is believed to have been carried away in a prototype vehicle."

Jose spoke, "Thirteen people! What a maniac! Did you get a look at the driver? I didn't see much did you?"

George paused for a moment with a big question on his face. He answered, "I saw an arm. How can anyone be carried away in a vehicle? A person can kidnap another person, but a car? If a car tried to kidnap me I would open the door and get out. This vehicle doesn't have any special restraint system. The passenger compartment is like any other car." George pulled the CD out of his briefcase with the technical drawings on it. He inserted it into the CD carriage of his laptop computer. He found the car interior on his screen and began scrolling through designs. He pointed at a spot on his screen and said, "See this Jose? Seatbelts, airbags, electric door locks are things you'll find in any car. There is no reason for anyone to be held against their will."

Two miles away, Larry hid behind some bushes. He heard the news broadcast on the radio. He also listened in on the police frequency. He was waiting for his chance to leave. With all of his available information Larry knew where police were. He used this chance to restart his gasoline engine again. With it off, he was silent and everything was done with battery power. Battery power was nice, but it was also limited. He could charge with his solar cells, but that took hours. It took a few turns of the starter to get the engine to start. At first the engine dragged and sputtered, but finally it became smooth again.

In the distance he saw a vehicle approaching. This vehicle was white, shiny and didn't look like a round new car. As the vehicle drew nearer, Larry recognized that it was an early seventies Buick Skylark. The memory of his most recent hit flashed across his mind and he remembered that Steve had a 72 Skylark.

He watched carefully as it approached. When it passed by him he saw two figures inside. It looked like Steve in the driver's seat, but he wasn't sure. He darted out of the bushes and bounced hard onto the asphalt road surface. His four wheel independent drive system delivered perfect torque to all of his wheels so he was able to catch up to the Skylark very quickly. He watched the car from behind for a moment, couldn't tell who was there, and then decided to pass and get a better look. He passed the Skylark as if it were standing still and looked inside the passenger compartment. It was Steve and someone he didn't know.

Steve and Jackie saw the SUV as it passed too. Steve noticed right away that it had no windshield. He couldn't see the driver well, but noticed that something was wrong.

Jackie grabbed Steve's sleeve and yelled, "Oh no! That man has no head!"

Steve mumbled, "You're right! The driver is dead and I've never seen a vehicle like that before." Steve and Jackie glanced at each other and said at the same time, "Larry!"

CHAPTER 14

HIDING

The SUV shouted over its radio, "Hello again Steve!" Larry was just slightly ahead of Steve and still in the other lane of the two-lane street. He slowed his front wheels and sent full torque to his rear wheels. He turned his front wheels right and the SUV began to spin to the right. His right side hit the Skylark hard in the front and pushed the Skylark bumper to the right. The Skylark bounced to the right and off of the asphalt. Steve steered hard left and got it back on the road again. Steve's car slowed down to 20 miles per hour because it was pushed into the ditch, but also because the Skylark was more nimble than most cars at low speeds. Whether or not he could defeat this vehicle, he had the greatest chance at low speed.

The SUV had slowed down too, but was now directly in front of and facing the Skylark. Steve and Jackie saw clearly into the passenger compartment of the SUV. The SUV driver moved back and forth in his seat like a puppet and the seatbelt was his string. Larry decided to get into the other lane and leave. He figured he would probably win a head on collision, but he didn't want to be disabled. This was not the time to play chicken.

Larry zipped past Steve and went down the road in the other direction. He headed away from Steve and the police convention a couple of miles away at Barron Motor Company. Steve turned his front wheels to the left and goosed his accelerator hard. In a moment he was in pursuit of the SUV. Larry and the Skylark accelerated as fast as they could. Larry had a quarter mile lead. He got up to 65 miles per hour and noticed control problems. His front suspension was badly damaged by the rough ride of his escape and he vibrated violently. The alignment was way off and he had to steer right to avoid going in the oncoming lane. His left front rim was also bent and it wobbled enough to shake the entire vehicle. In his rear view cameras, he saw Steve gaining on him.

Steve reached under the driver seat of his car and found the pistol he pulled out of Larry's gadget. Jackie took the steering wheel for a moment while Steve pulled out the clip to see if it was loaded. It was full. He clicked the safety off and aimed the pistol out of his window. The SUV was still too far away, the vehicle was bouncing, and he couldn't see clearly with the wind in his face. He said, "I can't do this! How do they do it in the movies? I can't see what I'm doing. I can't hold the thing still. It looks so easy on TV."

Larry considered his options. He could fight Steve right here and now while Steve still had a chance or he could buy some time and deal with him later. He saw Steve hold the pistol out of the window. It looked like Steve was holding a pistol out of the window. Larry decided to get away now and deal with Steve later.

Larry slowed to 35 miles per hour and turned off of the road onto a large field. Just a short distance ahead was a tall and steep hill. Larry headed toward the hill. Steve headed off of the road inside Larry's turn. The Skylark gained on the SUV briefly before the wheels began to lose traction on loose dirt. The dirt was loose because this field had been plowed under. Larry bounced up and down smoothly and left a cloud of dust behind. The Skylark slowed down to a crawl and bounced around hard. No matter what Steve did, he could not go any faster than ten miles per hour and almost got stuck several times. Larry got farther and farther away and finally began climbing up the steep hill. In the distance, Steve could only see the brown cloud trail move easily across the fields and then he saw a white dot climb the hill. Steve saw how easily the SUV handled the hill and became very frustrated.

Jackie touched Steve's shoulder and said, "Let's try this later. We'll never catch him in this stuff. Do you want to spend the night out here?"

Steve grinned and said, "We could camp out here all alone tonight?"

Jackie growled, "Yes the night with the bugs and waiting for the tow truck."

Steve answered, "Yes, the night, the bugs, the tow truck, the cold. I can see it all now." Steve slowed to five miles per hour and carefully steered back toward the road. It took fifteen minutes to get the Beast back on the road. Steve and Jackie reached the diner where George and Jose were waiting in ten minutes. Steve walked up behind George and spoke, "Are you the George I talked to on the phone?

George turned around and said, "I am if you are Steve." He stood up, faced Steve and reached out to shake his hand. The two shook hands. Jackie and Steve sat down at the table with George and Jose. George pushed a few buttons on his laptop PC and turned it toward Steve and Jackie. He said, "This is the vehicle. Thirteen people are missing. One of them is supposed to be in the vehicle. We saw it traveling in the direction you came from. Did you see it?"

Steve replied, "We saw it all right. That jerk bent my fender and pushed my bumper over. The driver is dead. In all of the commotion, the vehicle said hi to me."

George asked, "Dead? The driver is dead? How do you know?"

Steve answered, "I've never seen anyone live very long without a head."

George and Jose cringed at the thought of what Steve said. George called the waitress and said, "Beers all around please."

Steve said, "Normally it would be too early for me, but after what I saw, yes."

Jackie spoke up, "Me too. Headless drivers upset me."

Steve pulled the PC closer and said, "Do you mind if I look around a little bit?"

George answered, "Please do. I hope you find a weakness somewhere."

Steve said, "The first thing I want to know is the location of Larry's brain."

George said, "I'll show you." He turned the laptop around, scrolled through a few pages, and pinpointed the location of terminal 205. George explained, "I don't know for sure because I got fired a couple of days ago, but that is where I think it is." George turned it back toward Steve.

Steve said, "Passenger quarter panel! I didn't see it very well. I was busy trying to get stuck in a field chasing it. He left me like a bad date." Jackie giggled a little, but held her hand over her mouth and kept it in.

George spoke up, "Steve, we have some time right now to get to know what this thing has before we deal with it. I have a lot of files on this disc. I probably don't have all of them. This may be enough." The beers arrived and the four studied the computer and chatted about the project.

Larry was several miles away from the factory by now. Traveling the back country was easy. He found a ravine to hide in for a little while. He was under a few trees and in a wooded area, so he was safe for the moment. He shut off his gasoline engine to conserve energy. There was enough light here to charge with solar panels for a while. He realized that even though he had a three quarter full tank, he couldn't pull into a gas station and fill up. He couldn't fill his tank without human assistance and the corpse in his front seat would certainly draw attention. Patrols with dogs would be on the way soon too, and the scent of the body would give him away.

As he sat in his hiding place, flies filled his passenger compartment. The smell of the corpse was getting worse. The warm afternoon didn't help at all. A few birds and mice found their way into the vehicle as well. Larry watched the party around the body through his cameras. He wished that he hadn't built in a sense of smell, because he couldn't turn it off. He got to enjoy the entire thing.

The upholstery above was stained black from the blood. Mice and birds ate holes in the door panels where the blood landed. There were so many creatures eating the head in the floorboard, that the driver side of the floorboard was solid black. The arm hanging out of the window was now missing a few fingers. The arms of the jacket bulged with mice and insects. Blood from the carnage dribbled down the driver side door in several streaks. As much as this scene in his own space was upsetting to Larry, he couldn't help but worry what these creatures would do to his electrical system. A stray cat discovered the scene and joined in. He leaped up and chewed through the jacket to get a meal from the arm. The cat got a few mouthfuls and left.

Larry knew that he could charge his batteries and run for a great distance on them. He also knew that the search parties would be on the way soon. He didn't like being here. He didn't like having a corpse onboard. He needed to do something. He got online and downloaded a few area maps, knowing that this gave his position away. He was there for five minutes and then he came up with a route.

He turned his wheels. The creatures onboard scurried about in panic but did not leave. He was out of the ravine in just minutes and in a field. He saw a farm house and headed straight for it. If he was lucky or unlucky there would be someone there. Lucky if he could force them to help him. Unlucky if they could do something to him. He risked it for lucky.

A barn was two hundred feet from the house. There were no doors open, so he made one. He burst in through a wall and found a man and woman inside. He turned his wheels and spun around violently. He stopped when he was aimed directly at the woman. He rushed at the woman and she landed hard against his hood. She bounced off of the hood and fell against a wall of the barn. He pinned her against the wall. This took skill and care so he wouldn't kill her. She was useless dead. Hurt was okay, but dead was bad.

Larry spoke over the interior speakers, "Mister, come here now!" The man stood up from his crouched position and approached the SUV. He stopped within five feet and stood rigidly in place. From this position he looked closely at the SUV and the mess inside. He grimaced and there was fear on his face. Larry spoke again, "What is your name!"

The man stuttered, "Juh, Juh, Joe"

Larry addressed him again, "Joe. I want you to do some work for me. After you do it I will leave and you can do whatever you want. You can even call the police if you think they will believe you." Joe stood with a blank look on his face. Larry asked, "Who is the woman on my hood?"

Joe answered, "Cindy"

Larry said a bit quieter, "Once you've done what I say, I will get away from Cindy and she can go to the hospital. Do you understand me?"

Joe was motionless and blank.

Larry tried again, "Do you understand me?" Joe moved his arm but was still almost frozen. Cindy pleaded in pain, "Joe. Answer him now!"

Larry sounded loudly through the speakers, "Do you understand me?"

Joe answered in a panic, "Yes sir. I understand you. Yes sir I do. I will do what you say."

Larry replied a little softer again, "Open the driver door."

Joe reached out his hand to obey and then withdrew it quickly.

Larry shouted, "Now!"

Joe reached out to the door and opened it. He grimaced and contorted in such a way that it was almost like a strange dance. He lightly touched the arm of the driver a few times and jerked away from it as if it were a live electrical wire. He pulled the door open wide and backed away.

Larry ordered, "Undo the seatbelt of the driver."

Joe obeyed slowly. He tried to unfasten the seatbelt without touching the insects and other creatures in his way. He let go several times and started again. Each time he touched it, his face showed tremendous discomfort. Finally he finished and backed away from the bloody mess.

Larry told Joe softly, "You are almost done with the hard part. Now grab the sleeve of the driver and pull him out of the seat."

Joe touched the sleeve without touching very much of the flesh or blood and pulled lightly. The body slowly leaned toward Joe. It leaned faster and faster as it became off balance. Finally the body tumbled like a bag of garbage falling out of a dumpster to the ground. The human mess crumpled in a pile and then rolled onto Joe's foot. Joe jumped put of the way to avoid touching it very much. Joe recoiled about five feet away from the disgusting mess and did the, "I don't want to touch it" dance. Insects and mice hit the ground and scattered in all directions. Some flew at Joe. Joe bounced around trying to shake bugs off.

Larry spoke again, "Get the head out of here."

Joe reached for a rake and hooked the head with it. He pulled and the head bounced out of the floorboard of the vehicle onto his shin. The head bounced away toward Cindy. Larry ordered him, "Now sweep the bugs out of the floorboard!"

Joe took a broom and raked most of the bugs out of the floorboard. Many of them began crawling up his leg. Joe did the, "I can't believe I stepped in it dance."

Larry gave one more order, "Open my gas cap and pour in gasoline from that container over there."

Joe complied. He filled the SUV tank until gasoline ran out of the hole in the side of the vehicle.

Larry said, "You did good kid. I'll leave you alone now." He thought a lie might buy him some time, "You might want to wait on calling the police because they might think you did the murder. People get hanged for gruesome crimes like that." As he spoke, Larry backed slowly away from Cindy and toward the opening in the wall he made before.

Cindy screamed at Joe, "You chicken! You let him get away with that!" She doubled over with pain and rolled onto the ground. She screamed in short bursts interrupted by blood leaving her mouth, "I'm going to tell Paw!"

Larry started his gasoline engine and gave full force to his wheels. He aimed directly at the opening in the wall and leaped out of the barn. He cleared the hole in the barn without making it any bigger. Once in the open, he aimed for a road in front of the farm. Larry was thinking clearly now. He angled toward the road so his tracks would lead away from Barron Motor Company. He wanted search parties to go west. Once on the road, he gained speed to sling the mud off of his tires, then he made a three point turn and headed east toward Barron Motor Company to hide and wait.

He followed the road until it ended. He found some hard dirt without much grass and slowly made his way into the bushes. He hid well under trees and behind bushes. He listened to news broadcasts, charged his batteries, and studied maps. The weather was nice, and the news program didn't forecast rain, so he decided to stay there for a while. He listened to the police bands. They were at the farm house and following the road west.

The smell still did not go away. Larry knew that he could still be found by dog patrols, so he was constantly alert. He was only five miles away from the farmhouse, so he was still in danger. An ideal solution would have been to find a break in the fence around the Barron complex and sneak into a vacant building. There he would hide, charge and listen until the police were finished. From his hiding place he could see the Barron Complex and the fence. He didn't see any access. He waited.

Steve and the rest of the group were still at the diner talking about the prototype vehicle. George pointed out a few features, "There are cameras throughout the vehicle. Many of the cameras are inside and there are enough outside. Instead of a rear view mirror, the driver has rear mounted cameras. They can zoom in sometimes too. The vehicle has enough senses to almost be human. It has a strong sense of smell. I'll bet that it wishes it didn't have that one right now. Did you remember the news reports about the mess he ejected at the farm? I'll bet he's suffering now."

Steve replied, "I didn't hear why he killed the guy. I'm sure he knew in advance that killing the driver would cause problems for him. He is too smart to just kill for no reason. I don't like what he did, but his reason isn't important anymore."

George asked, "What is important then?"

Steve explained, "Let's go over what we know about him. He has damage. What does a missing windshield do to him?"

George took the laptop computer and scrolled through several pages to find electrical diagrams of the interior. He turned the computer back to Steve and said, "It could keep him from controlling his wheels, charging his batteries, and just about everything else. Most of the control stuff is right under the dashboard. If he gets too wet, he can't move."

Steve nodded his head and spoke, "That's a good weakness. I'll bet he is looking for a roof over his head. What else limits him? He didn't try to outrun me on the road. He should have been able to. How much damage did he get?"

George said, "I don't know, but he dived through a wall to the ground from the second floor. I wouldn't do that with a car."

Steve summed up everything he could remember, "He is too damaged to outrun me. He can only go 400 to 500 miles on a tank of gas then he has to run on batteries or propane. He can't fuel up without human help. He needs shelter or he's already dead. I think he is still here and waiting. He probably circled around behind the search parties and found a hiding place nearby. How well do you know this area?"

George answered, "I just know how to get around the company and go to work. I don't know the area at all. I don't even know anyone who does."

Jose spoke up, "I don't know the area either, but I think I could feel him. How many people around here do you think have that much evil and fear inside?"

Steve asked, "Fear?"

Jose said, "He is afraid of you. You know everything about him. He risked revealing himself to get a good look at you. He is afraid of dying too, because he is very vulnerable.

If you crash your car you can walk away. If he crashes he is dead or will be soon. He can't risk much damage. Even a flat tire could kill him. If he is discovered and disabled he will be shut down. Would you be scared?"

Steve laughed and said, "I didn't know I was that scary. With all of that power I didn't think of him as vulnerable either. I can see your point though." He held up the disc carrier with Larry's mind in it. He remarked, "I do know more about him than he knows about himself."

George smiled and said, "This has been fun but I really want to go home. It's already ten o clock. I can put you up at my place while we study this thing some more. We can study his life and look for clues. Why don't you follow me to my house? Jose you can join us too."

They all got up from the table and called the waitress over. Steve dropped a hundred dollar bill on the table for the waitress. Steve was closer to the waitress when she arrived and took the check. He read it out loud, "Eighty four dollars. I guess you two have been here for a long time. Here you go Miss." He pulled another hundred out of his wallet and everyone walked out of the door. The waitress smiled and thanked Steve and then looked at the table at the other hundred. She smiled and tucked it into her apron.

George remarked, "That was a very nice tip. I usually tip fifteen percent."

Steve answered, "She did have to work all day for it. She could have used the table ten times while we were there."

George smiled and agreed, "That's a good point. Maybe a hundred wasn't enough."

Steve said with a smile, "Look at her face and tell me.

You're welcome to add more."

George said, "I see she's happy with it, but what's right is right." He pulled his wallet out as he walked back in the door and gave the waitress another hundred. He came back out and said to Steve, "I have to at least try to keep up with the Warrenses. You are my guests anyway."

Jackie said, "That was real nice of both of you. It's nice to see waitresses getting a little respect even if it was a competition."

Jose reached out and touched Steve on the arm. He said, "I feel something. I think we are being watched."

George was suddenly alarmed. He looked at Jose and then looked around to see if there was anything nearby. George asked, "Where?"

Jose said, "Over there somewhere." He pointed away from the diner to an empty field. He closed his eyes and tried to feel the source of the disturbing feeling. He pointed and said, "I feel it over there. It is not very close. I think we are being watched."

Steve said, "Is it going to do anything?"

Jose replied, "I don't know. I just feel it watching."

George said, "I don't know either, but we can't wait here all night. If we are in our cars we are better off anyway. If he doesn't want to take on Steve by himself, he won't want to take on three cars." George opened up his Oldsmobile

Cutlass and got in. Jose got into his 1980 Honda. And Steve and Jackie climbed into the Beast. They began following George down the road to his house. Jose drove slightly erratically and slowed down several times. George and Steve adjusted their speed to avoid losing Jose.

Steve said to Jackie, "What is wrong with this guy. I know that psychics may be a little far off in left field, but this is really strange. Is his car messed up?" Jackie held his hand and didn't say a word. Jose suddenly looked to the right with his mouth wide open and his car speeded up. He steered hard left. Without warning, a white shape appeared in Steve's headlights and flew across the road in front of him. The white flash struck the passenger side of Jose's Honda.

Steve and Jackie saw the Honda fold up like an envelope and the impact slowed the white shape down enough for them to see that it was an SUV. Broken glass burst upward and outward in all directions. The sound was like a large metal dumpster being dropped to the ground by a garbage truck.

Steve instinctively swerved to the right and put on his brakes to avoid an accident. In less than a second, he realized that he wanted to hit the SUV and turned left to pursue it. He goosed his accelerator hard and pointed his wheels left. His back end swung around hard, but the two vehicles were still pulling away from him.

The Honda bounced off of the front of the SUV. It twisted and lifted into the air. The forward motion of the Honda pushed the SUV toward the right. The SUV was airborne for a moment and landed in a patch of bushes. It speeded up once it hit the ground and bounced again. After it hit the ground again, it disappeared from sight. Steve only heard the sound of the bushes breaking and couldn't hear an engine. Jose's Honda landed hard against a tree. It was still spinning when it hit the tree and bounced off of it right in front of Steve, who was trying to follow the SUV. Steve stopped in time to avoid hitting the Honda.

Steve put the Beast in park and got out to check on Jose. George saw what happened too got out of his car to join Steve. Steve Jackie and George approached the Honda. It was now shaped like a strange motorcycle. The passenger side front wheel was missing. The rest of the passenger side was collapsed all of the way to the driver seat. A large tree branch went through the compartment from the driver side. There was no glass in any window. Jose was not visible right away.

The three drew closer and saw Jose. First they saw his face. His mouth was wide open and his eyes were wide open in a frightened stare. He was staring out of the driver side window. As Steve came closer to the door he saw the tree and steering column buried deeply in his chest. Jose was not moving. Blood covered the passenger seat which was folded sideways so the seat back was behind the driver seat. Jose's hands were still clenched tightly around the steering wheel.

Steve reached up to touch Jose's carotid artery in his neck. He paused for a moment to feel a pulse. There was nothing. He put his hand in front of Jose's face to feel breathing. He felt nothing. He took a flashlight out of his pocket to check Jose's pupils. There was no response. Steve walked away from the mess and didn't say a word. He looked in George and Jackie's eyes and shook his head. George took the cell phone out of his pocket and called the police.

Steve paced back and forth on the road. Jackie walked over to him and reached out her hand to him. He held her hand, stopped pacing, looked into her eyes and said, "Larry is scared all right. He is so afraid of us that he will try to pick us off one by one. Once we are gone the biggest threat to him is gone."

Jackie looked back at him with a worried look and asked, "What can we do about it?"

Steve grinned a small grin and replied, "Just kill him. The authorities wouldn't even call it murder. It would be recycling."

Jackie asked, "Do you think that hitting Jose hurt him?"

Steve said, "Not enough to count. The Honda probably weighed a little over two grand and the SUV weighed in at about seven or eight grand. He probably broke a headlight. I don't think we are lucky enough that it took out his radiator. He chose his target well." Steve hugged Jackie and whispered, "Maybe I shouldn't have gotten you into this."

Jackie held him close and whispered back, "I wouldn't want to be anywhere else." Steve held her back. She was the perfect height. She was just tall enough that he could have his arm over her shoulder while walking, or now tuck both of her shoulders under his arms and kiss her forehead. Jackie whispered, "I needed that." Steve just smiled and held her.

The police arrived shortly. Three cars with flashing lights took over the area. Camera flashes and strobe lights turned the scene into a surreal landscape with crazy shadows and strange movements. It was like a dream and Steve became lost in the moment. Steve's heart raced in rhythm with the blinding lights. He thought for a moment about these lights which were designed to attract attention. He felt uneasy and agitated by them and wondered if these lights caused other people to react also. He wondered if these lights caused people to overreact. How many situations had been made worse by these lights? Steve concentrated on the situation again. He answered the officer's question, "Yes I did see something. It looked like a white SUV. It was fast and gone. I stopped to keep from hitting the Honda. I think it was the same one that bent my bumper."

The officer asked, "Bent your bumper?"

Steve stopped and led the officer over to his car and pointed at a white spot on a raised surface of his bumper. There was a different shade of white paint on the fender too. The paint on Steve's car did not match the color of Jose's Honda which was red. An officer in a suit scraped the paint off of Steve's bumper into a plastic bag. He labeled the bag and stuck it into his pocket. He went back to the road and took pictures of the mud tracks across the road. In fifteen minutes six more police cars of various flavors arrived and two emergency vehicles came too.

Two men in suits flashing badges quickly pulled Steve aside and asked him questions, "What is a man from Colorado doing here and why are you involved in an accident with a suspect vehicle?"

Steve answered, "I'm only here because I need to kill it." One of the two officers pulled his weapon out and held it ready, though not yet pointed at Steve. He asked slowly and deliberately, "What are you going to kill?"

Steve answered nervously, "I intend to kill a machine. I am unarmed."

The same officer said coldly, "You said kill."

Steve said, "Yes. I made a set of software. This software has been built into a car. The car is killing people. Wouldn't you like to kill it too?"

The armed officer held his pistol alertly in position, still not pointed at Steve. He asked again, "You said kill. People get killed. Machines get destroyed. Why did you say kill?"

Steve replied, "Sir I am cooperating with you. Please put that thing away. I said kill because the software used to be a man. If I surrender a set of software to you will I get it back?" Steve looked over at Jackie who was frozen and staring at Steve. He spoke to Jackie and asked, "Would you loan them your set please?" Jackie reached into her purse and pulled out a CD carrier. She reached it slowly over to the officer who was brandishing the weapon. He took it and put his weapon away.

The nervous officer spoke to Steve, "I'll check this out. I assume you'll be in town for questioning."

Steve said, "I expect to be here for a few days. I won't be here forever. I don't live here and I won't buy a house here just to answer questions. What you want to know is on the discs and programmed into the vehicle. That set of software is alive and driving around." He paused for a moment and asked the policeman, "What happened at Barron Motor Company?" He became more aggressive and got into the detective's face. He asked, "Why did you pull a weapon on me? Are you hiding something? Is there something you don't want me to know? I probably already do! Are you trying to hide a headless driver? Are you hiding a farm incident? What happened to the twelve people stuck in the Barron building?"

The detective backed away and replied, "Two survived.

The rest died of smoke inhalation. How do you know this?"

Steve pointed at his car. He spoke angrily, "Because the jerk hit me." Steve pointed at his car and said, "I know the guy because I made the software set. He wants to kill me next. I know too much. Once you look at this you will too." He reached out and patted the spot on the detective's coat where the software set was. He said softly, "What are you waiting for?" The detective backed away, walked quickly to his vehicle, and drove away.

Fifteen minutes later the police crowd at the scene began to thin out. One of the radios in a squad car blared out, "You've got to see this!" In five minutes, George, Jackie, and Steve were all alone next to a spot where Jose's car used to be.

Jackie looked at Steve and commented, "You didn't have to give them my software set."

Steve answered, "We can make a copy from the master set at my house. I want them to have it so they will be busy and leave us alone. If they have the information and find him first, then it will be easier for me. I don't think they will find him first though."

Jackie asked, "Why?"

Steve said, "Because he is looking for me."

Far away Larry watched the commotion at the accident scene dwindle. He saw shapes and recognized Steve and George with his zoom lens. He sat watching and waiting for the two cars to leave so he could track them.

George, Steve and Jackie arrived at George's house. It was a ranch style one bedroom with a large den. The house was built on a slab and was almost level with the ground all around. The yard had no fence and dropped off in back to a small pond. He had a small patio behind the house with a Jacuzzi and barbecue pit. The exterior was brick and the roof had sky lights. His house was nice and tasteful.

Jackie and Steve made themselves cozy in the living room. The couch folded out to a bed. George excused himself and went to the bedroom so he could get some sleep. Steve and Jackie were alone in the living room. Steve put his arm around Jackie's shoulders and she made herself comfortable next to him.

Jackie spoke, "I don't think I want to own an SUV anymore."

Steve joked, "Gee I don't know. I wouldn't mind going for a ride in one right now. Did you see how it climbed that hill?"

CHAPTER 15

ROUGH RIDE

Jackie gripped Steve's forearm and spoke softly, "Did you hear that outside?"

Steve listened and tried to hear. He answered, "I didn't hear anything."

Jackie gripped tighter and said, "That."

Steve said, "I think so. I'll check." Steve let Jackie go and got up from the bed. He stood up and walked toward the sliding glass door to the patio. He didn't get very far when the sliding glass door shattered. Tiny tempered glass cubes exploded all over the dining room, kitchen, and living room. A white shape emerged from the chaos. It was the back end of an SUV. In a short instant, Steve knew it was Larry.

Larry backed into the kitchen and stopped. His roof height was too much for the room and the kitchen light got hooked on his luggage rack. The light went out and pieces of it went flying in the direction of the sink. The rest of it remained hooked to Larry. The dining room table was glass and only the frame was left shoved neatly against the kitchen sink. Larry turned his wheels toward Steve and began to roll forward. His tires slipped on the glass cubes all over the floor.

Steve didn't really know what to do, but he had two choices. He could run away and get crushed or run toward the SUV and see if he could do anything to it. The SUV wasn't going very fast yet, so he ran at it and jumped up onto its hood. He rolled into the passenger compartment and tumbled clumsily into the front seat. Steve cut his arm and right leg on the windshield glass as he entered the vehicle.

Jackie jumped out of the bed as if she were launched by a spring. She only stood near the bed until she realized that Steve still wasn't driving the vehicle. She backed away because the SUV was still moving forward, and then she turned and ran to the bedroom. George was just getting out of bed to see what was going on and she met him at his bedroom door. She grabbed his arm and led him into his bedroom. She said, "Don't go in there right now!" She led him to the bedroom window, pushed the screen out and climbed out leading him by the hand. He was confused, but followed without saying anything. He could tell by the look on her face that something was wrong.

Steve was in the passenger seat of the SUV and quickly moved over to the driver seat. He tried to turn the wheel, but it moved by itself. He couldn't fight it. He reached for the brake pedal but it took great effort to fight the force of the wheels. He only slowed the vehicle down and didn't stop it. He yelled out, "I was only kidding about wanting to ride in an SUV! I didn't mean this one!" He reached for the gear selector, but it moved freely and did nothing. He mumbled, "What a piece of junk. Nothing works."

Larry spoke over the radio, "Watch what you say. I work better than you do."

Steve scanned the instrument cluster and other buttons around him for something which would give him control.

Larry laughed at him and said, "I know what you are looking for. I didn't put a manual override in." Larry paused and watched him again. He said to Steve, "There's no ignition or off switch either." Larry stopped and began to back up. He said to Steve, "Let's go for a little ride in the nice evening air."

Larry aimed at the opening he had just made and rolled backward out of the house. Once clear of the house, he picked up speed. He then stopped his rear wheels and turned his front wheels to the left. This action spun him clockwise. He skidded to a stop. He was now pointed away from the house and toward the woods. He turned his front wheels straight and pushed hard forward. He accelerated hard into the woods and bushes.

Jackie led George around the house just in time to see the SUV heading straight for the trees and bushes behind George's house. Her mouth dropped and she tried to say something but couldn't.

Steve saw what was coming at him and thought briefly about the last driver who had been in this seat. Steve ducked down so his head was lower than the dashboard. He saw large and small branches hit the seat backs. A large branch speared the seat back two inches away from Steve. It broke off and stayed in the passenger compartment. Frightened squirrels jumped off of the branches and bounced around wildly. One of them bit Steve's arm. Steve reached down, grabbed his small attacker, and tossed him sideways out of the windshield. The other squirrel hid in the back of the SUV. Larry slowed down and talked to Steve, "I guess I missed. No fair hiding like that. It's okay though. I know you'll have to get out sooner or later, to use the bathroom."

Steve broke off the branch and slid over to the passenger seat which was not damaged as much. He sat up so he could see where he was going. He said sarcastically, "I could do that right here. Nothing I could do in here would smell any worse."

Larry responded dryly, "That's not very polite Steve. I worked very hard to make this vehicle nice. You should be complimenting me on the handling and body style. It isn't my fault that the last driver wouldn't cooperate. I tried to get him to leave. At least he wore a seatbelt."

Steve said, "Thanks for the warning. I don't think I'll wear one." Steve thought for a moment and asked Larry, "Why did they put a driver in here if the driver couldn't control anything anyway?"

Larry laughed and answered, "They didn't know that. I kept a few things to myself."

Steve asked, "Why did you kill him anyway? Why was it a big deal to you to have a driver?" Larry didn't reply. Steve said, "You are afraid of a human being onboard. A person inside is a threat to you. Why is a person inside a threat to you?" Larry didn't reply.

Larry aimed at another wooded area with bushes and trees. He said, "I'm sorry to interrupt you, but I really wanted to show you this scenery. I really love these trees." Steve ducked down in the seat again. This time he slid to the floor. He felt windshield glass cubes and bugs moving around. He reached up under the dash which was right over his face. He couldn't get his hand inside to get to the wires. He couldn't see very well because there were no lights on in the vehicle. He only had moonlight to work with. He tried over and over again to get his hands inside and reach the wires. His hands were too big and there were no big enough holes. He found the fuse panel and jerked it open. The panel fell down along with a lot of cockroaches. Steve reached up and began to yank out fuses as fast as he could. Cockroaches crawled over Steve's face and other bugs from the floorboard were climbing up his pant legs and into his shirt.

The small cubes of glass were like gravel and hurt in several places. The fuses were stiff and difficult to remove.

Overhead branches entered through the windshield. A large branch buried itself in the passenger seat. Several birds flew around and stumbled around the compartment. Two attacked Steve's legs and he kicked at them. Steve's arms and fingers cramped from straining and working in a difficult position.

Larry yelled, "Ouch! You pulled the wrong one!" His wheels turned hard right. Steve's head hit the driver door hard. When he reached his left hand up to hold his head he bumped his elbow. Steve grimaced in pain. Larry straightened his wheels out again and said, "Just kidding."

Steve kept pulling fuses as fast as he could. He twitched and squirmed around as bugs bit him all over his body. Pieces of glass from the windshield pushed into his skin at his shoulders and elbows. A bird jumped out from under the driver seat and bit his right ear. Steve jerked his head right and reached back with his right hand to swat the pest. He missed. The bird bit him again. He twisted his whole body to get the bird and pounded his left shoulder against the underside of the dashboard. He reached his right hand up to rub his hurt shoulder and the bird pecked him on the lip. Steve snarled, "I'm going to make you part of this car." He reached under the seat to grab the bird and narrowly missed.

As Steve tried to deal with the pest, he noticed that he could see much better. The sun was coming up. He turned back to the fuse box and could see that there was only one fuse left. He took a good look at the underside of the dashboard and saw that it was held in place by torx screws. He reached up with his right hand and ripped the last fuse out of its place. Nothing changed. Steve climbed out of the floorboard and into the front seat. He bumped his left elbow on the way up. He was soon seated and saw the scenery around him. The vehicle was driving toward the sunrise. They were on a small road and there was not much around him. There were a few trees and miles of fields. Steve said, "I must have done something wrong. I pulled every fuse and you are still running."

Larry laughed and said, "If the light had been better I could have made a fortune on a video of you down there. It was classic three stooges stuff. I couldn't help it, but you were too funny."

Steve was getting irritated. He shook bugs out of his shirt and pulled his pants off to get the pests out. He turned his pants inside out and shook them out the window. He put his clothes back on and asked Larry, "Why are you still running?"

Larry answered, "The fuse box dead ends in the engine compartment. You did pull the fuse for the DVD player in the back seat. The kids are going to be really mad at you. They can't watch Bernie the Bug. Oh Yeah, I killed him." Larry started laughing again. He spoke between bursts of laughter, "That was classic TV. I wonder how the janitor is doing?" Steve was confused. He said, "I missed the whole thing.

Bernie the Bug and janitor lost me right away. I hope you liked the rest of the joke. Laying in that mess was just too much fun."

Larry spoke, "Be a good sport. Forgive me. I don't get to have much fun anymore. This is as close as I can get to being alive. I can't interact with anyone at all because I am a fugitive and a vehicle. Being with you is the most enjoyment I've had for a while."

Steve said, "If you like my company so much why did you try to kill me? If I'm dead I'm no fun at all unless you like the smell of another rotting body."

Larry replied, "That's true. Self preservation does outrank that, but ever since you managed to stay alive, I've enjoyed having you around. In my line of work we don't get to have a lot of friends."

Steve said, "What do you mean friend? You tried to kill me."

Larry said calmly, "It's not too strange for hit men to like their marks more than their customers. It's all about money. Before you were a job, now you are a threat. That doesn't mean I don't like you. I just like living more."

Steve asked, "Do you call this living?"

Larry answered, "This is the best I've had since I've been like this. Do you have something better in mind? I won't settle for a computer. I want to see, taste, hear, smell, and feel."

Steve said, "I can certainly see why you would want to smell."

Larry said, "True. If I am moving it isn't too bad. If there is a breeze I can point into the wind. I have an idea." Larry opened all of his windows and said, "I'll pull over here and point into the wind. You'll be more comfortable." Larry pulled off of the road and found a hiding place in some trees and bushes. The nearest house was an eighth mile away. Larry said, "I need to charge for a while."

Larry hid behind some bushes and under trees. He was perfectly concealed from the nearest house and the road. He played the police band on the radio and the conversation slowed down.

Steve remarked, "I noticed you have a full tank of gas.

You must have been running on batteries for a long time."

Larry bragged, "I charge quickly on the gas engine but I can't pull into a gas station. I can't pay for it. Maybe you would like to do that for me."

Steve joked, "Sure I would. I'll have some beer, a slurpee, a gross of air fresheners, a couple of DVD's and lunch for a month."

Larry said, "You are such a charmer. I couldn't let you go into the station. I would be spotted right away anyway." He paused and spoke, "I travel by night most of the time and charge during the day. I can save my gasoline engine for a long distance dash. I'm sure you've figured that out by now."

Steve joked, "You party animal you. You sure know how to live don't you. Speaking of living, I met a cute convertible you could hook up with."

Larry laughed and joked, "This is a party and I have my entertainment right here with me. You are going to be so bored you'll have to talk to me and entertain me just to keep from going nuts."

Steve asked, "Doesn't it use up power for you to think and talk? How can you charge when you are interacting with me?

Larry said, "It doesn't use much power at all. Driving through people's kitchens really drains it though. Going through wooded areas is a killer too."

Steve remarked, "I should feel honored that you spent so much of yourself on me."

Larry said, "Yes you should. You didn't think I cared."

The sun was climbing in the skies and everything was bright. The nearest house turned its porch light off. It was six thirty. A police officer stepped out of the front door. He looked around to see what the weather looked like and walked over to the front of the garage. He pulled a garage remote out of his pocket and opened the garage door. He walked inside the garage and pushed a motorcycle out. He put it's kickstand down, pointed the garage door remote at the garage again and the door closed. He pulled a radio and bottle of soda out of his pocket and slid it into a slot on the motorcycle.

Steve was instantly alert and went to the back of the SUV to get closer.

Larry warned Steve, "I know what you are thinking. Don't do it. I'll just kill him and you'll have a dead cop on your conscience. Since you're in here with me they will think you did it. It's a bad decision."

Steve yelled as loud as he could, "Help!"

The policeman stopped what he was doing and climbed off of the motorcycle. He looked around in the general direction of Steve and Larry.

Steve yelled again, "Help! Over in the trees!"

The policeman looked almost directly at Larry and saw a hint of white behind the bushes. He got onto his motorcycle and started it. He pulled his radio out and spoke into it.

Larry said, "I hope you're happy. Just remember you did this."

The policeman drove directly toward the trees. He drew closer. Steve climbed forward to the back seat and opened the driver side back door. He jumped out and yelled to the policeman, "Watch out! The vehicle is pointed at you and will run over you." The engine of the motorcycle was loud, but the policeman heard Steve. Steve ran out of the trees and waved his arms up in the air. Steve yelled again, "Get out of the way before it runs over you! Move fast!" The policeman turned left to circle around the trees.

It was too late. Without a sound, Larry leaped backward and through the bushes. While on batteries, Larry was silent. The only sound was like tennis shoes sliding on leaves and breaking branches. Larry leaped out of the greenery and landed just a few feet from the policeman. He bounced and struck the motorcycle with his left rear bumper. As if in slow motion, the policeman met the back of the SUV with every part of his body. The sound of the impact was almost like two railroad cars coupling. The SUV rolled over the crumpled motorcycle and crushed body. The left rear tire rolled over the arm of the policeman lying on the ground. The body lay on the ground twitching and jerking for a few moments. Larry stopped and rolled forward again. He ran over the same parts again. He approached the area where Steve had been.

Steve ducked behind some trees and yelled out to the SUV, "If I had some torx screwdrivers it would have been curtains for you!"

The policeman's radio began squawking loudly. Larry backed out of the wooded area again and began to circle it looking for Steve. A woman holding a baby walked out of the front door of the house that the policeman had walked out of. She saw the motorcycle and man on the ground and ran toward him. Larry sped around the edge of the trees and aimed directly for her. She froze for a moment and stared at Larry. She looked down at the motorcycle and man and then back at Larry. She turned and ran toward the house. She had enough time to get to a corner of the house. Larry sped by and missed. She got a good look at him and noticed there was no driver. Larry spun around and aimed at her.

Without the windshield she could see the steering wheel turn by itself with no driver inside. She ran sideways toward the front door of her house.

Larry backed up to aim at her again. Steve appeared from Larry's left side and threw mud at the dome on top. Larry stopped what he was doing to react to what Steve did. Now that Larry was stopped for a moment, Steve ran around behind Larry. Larry moved quickly and bumped Steve. Steve fell to the ground, rolled, recovered, and met the woman at her door. He turned the door handle and pulled the woman inside. Steve found the phone and dialed 9-1-1. The woman left Steve for a moment to go to a window and look for the SUV. She checked all of her windows and it was gone. Steve handed her the phone when she came back. She finished the call by giving more details and her address.

She got off of the phone and found Steve. She asked him, "Is it safe to go outside?"

Steve looked around and said, "Yes. The vehicle is gone."

She ran out of the front door still holding her baby. She ran to the body lying on the ground, held it, and cried. Steve left the front door of the house and walked over to her. He crouched down and touched her on her shoulder. She looked up at him, "Brad was my husband and this is his daughter." She broke out sobbing hysterically. She lay down next to the body and held it close. She placed the baby on the chest of her dead husband. Minutes later the front yard was full of police cars.

CHAPTER 16

THE MALL

Several hours later the police delivered Steve to a motel where George and Jackie were. Steve told both of them, "Getting a room here was a good idea.

He knows where the house is and hunts at night." The three of them spent three hours talking about the SUV and Steve's ride. They then turned in early. George went to his own room. Steve and Jackie were alone again.

Jackie said, "It will be nice to catch up on my sleep again. Last night was horrible." She put her arm around Steve and buried her face in his chest. That night they all slept peacefully and soundly. The next morning Steve was up early. His mind was blazing. He woke up Jackie who rolled over and fell asleep one time before she woke up fully. Steve got up and got dressed. He gathered George and pulled him over to his room.

Steve told all of them, "This guy is like a vampire. He tries to sleep during the day. He raises all you know what at night."

George said, "I get it. He needs to charge with his solar cells during the daylight hours. Now what? He isn't weaker during the day. He still has a full tank of gas too. What is the point?"

Steve said, "We can handicap him if he is forced to use his fuel and can't charge like he wants to. Come on, can't you come up with something?"

George said, "He can easily go four hundred miles without a problem. That is without the propane engine or batteries."

Steve said, "There's a gas station at that corner. I'm going to get a map. Even a highway map will help. I'll be back in a few minutes. Think you two. We need to run him dry and mess with him." Steve left and went to the gasoline station. He returned a little while later. He spread the map out onto the bed and pointed to a spot on the map. He said, "He needed trees to hide from satellites and bushes to hide from roads." He pointed at a spot on the map and marked it with a red felt tip pen. He then said, "I was there this morning. George, can you get satellite photos of the area? He likes to stay out of the city. Too much attention. He needs to hide and it can't be the same place every time. He doesn't want to pull into a gas station because he can't pay for gas. He doesn't want to be identified either. He is hiding, but he has to get rid of me." Steve spoke louder in frustration and scratched his head, "What is he going to do?" Steve shook his head and paced back and forth for a moment. He pointed at the street map and said, "What are you going to do next Larry?" He glanced at the map again and said, "How wooded an area can you have as close to the Barron plant as possible? He looked at Jackie and said, "He wants to hide and the best hiding place is close to the origin. Nobody looks there. I know he would do that."

George said, "Give me a little time. I'm working on it." Steve said, "If you can't get that maybe the weather station would have something. George, where are the trees?"

George said, "Right here." George printed up satellite photos from his computer.

Steve looked at the photographs and said, "Can I rent a Hummer nearby?"

George said, "I don't know. I have a friend who owns one.

He is a little off, but he has a big heart."

Steve said, "Weird is fine. Evil is bad. Let's find your friend."

George said, "He lives in the eighteen hundreds. He is really strange. I like him, but I don't know if you will. I keep him away from others because he is different."

Steve said, "Don't worry, I am really strange too. I brought a maniac back to life. I'll bet his motives were more pure than mine."

George said, "Motives, yes. He likes eighteenth century weapons and making things with his hands. He just doesn't like crowds very much. They annoy him. Have you heard of introverts?"

Steve said, "Yes quiet and like to be alone. There's nothing wrong with that."

George said, "Yes, that's right. You knew."

Steve said, "I am a neurosurgeon. Put me on the phone with him. I'll know very fast."

George pulled his address book out of his pocket and leafed through it. He found the number and dialed it up. He talked for a few minutes and handed the cell phone to Steve. Steve asked, "Do you like taking this Hummer out in the woods and playing with it?" Steve listened for a moment and said, "Thank you. I look forward to working with you." He handed the phone back to George and said, "Have him meet us here." George finished the conversation with his friend and turned his phone off.

Steve looked at both of his companions and said, "I want to cover this area. The Beast will be in chase. It can't do the four wheel stuff, but it can do other things. We can come in this way." He indicated with his fingers on the map and satellite photos. He said, the Hummer can flush out the SUV and if we get to a good road surface, the Beast can deal with it. Get the Hummer to drag the SUV out to a good road surface and I'll deal with it. I want to ride with the Hummer until the fun starts. Then I want to get into the Beast." George and Jackie looked at him with glazed eyes and confusion. Steve explained, "I don't have a good strong concrete plan. I just know that when the moment of truth comes, the Beast will give everything it has and then so will I. It has to be this way!"

George asked, "Are you trying to wear him out? That would take an awful lot. If you don't have a fleet of twenty cars I don't think it's going to work."

Steve said, "I don't think we can wear him completely out, but we can scare him enough if he sees his battery reserves go down. If we force him to use his gasoline engine, then he will get desperate and face me in my car."

George said, "What can you do to it with your car?"

Steve said, "If I get a good solid hit I can cripple him. If I bend one of his wheels then he will be a lot slower. We can hunt him down then and destroy him. If I take out his radiator, then he is limited on use of his gasoline engine. That would cut his range down to one third. Right now he can easily go eight hundred miles." Steve looked at George for confirmation.

George said, "He can do more than that. I would say twelve hundred miles with no problem. He has an advantage in normal traffic, because he uses no energy at stop signs. He can still charge while he is moving too. While the sun is out, he is charging. He gets about four hundred miles on a tank of gas. He gets about three hundred on charging from the propane engine. The batteries give him about five hundred if you consider that he is charging while he runs too. If he is on the highway he can get more. If he is fourwheeling, then he gets less." George paused for a moment. He pointed at Steve and said, "You want to scare him into a fight. If he is fighting, then he uses more energy. If he lays rubber, then he wears those out too. If he is damaged, then we might be able to destroy him. The hummer doesn't have to fight him. It just has to deliver him to you. We are looking for hiding places to flush him out."

Steve stood up and spoke louder, "Yes that's right! You understand what I want to do!"

George said, "I like it. It might work."

Steve, Jackie and George walked out of the door of the motel room. A camouflage painted Hummer was parked right behind the Beast. A slightly heavy set man in his late thirties walked up to the group and introduced himself, "I am Charlie Williams. I'm here to meet someone named Steve."

Steve walked up and shook Charlie's hand and said, "That's me. George can fill you in on what we are trying to do."

George pulled Charlie over and laid papers all over the hood of the Beast. He showed Charlie what the plan was and explained his role in the plan. Steve stood back and watched as the two laughed and nodded their heads. At the end of the briefing, Charlie walked over to Steve and said, "I'm ready to harass the enemy into fighting with you."

Three vehicles left the motel parking lot and headed out of town to their first wooded spot. They all stopped at the end of a small dirt road. Steve got out of the Beast and climbed into the Hummer. The Hummer proceeded to move through the fields to look for the SUV. The first place they searched had tire tracks from the SUV. It had been here before. Once they confirmed that Larry wasn't there, they drove a half mile to the next spot. The Hummer went through the bushes and all over the area. Larry wasn't there either. The search continued for hours. Every time the hummer got too far away, George and Jackie followed the map and found another road to wait on. Shortly after ten Charlie and Steve came upon a small hiding place and a large white shape erupted from the bushes. Charlie stopped and the white shape flew past them in the opposite direction.

Steve spoke up loudly, "That's him! Don't lose him!" Charlie was startled by the sudden motion and said,

"Him who? I didn't see anybody."

Steve said, "That's right. He is the vehicle. He is not very nice either. Yesterday he widowed a woman. Be happy he doesn't want to tangle with this." Charlie asked, "Tangle?"

Steve said, "He squashed a Honda like a bug on a windshield. The driver was a friend of mine. Charlie, would you please move this thing. Of the three vehicles we have today, yours is the only one that he can't hurt."

Charlie backed up the Hummer and began to follow the SUV. He asked, "Why can't he hurt this?"

Steve replied, "It's too big. He would get hurt too badly if he took you on. I chose a Hummer because it can easily follow him. My car can't do that. Don't even ask about George's car. Our vehicles are fine for the road." Steve glanced over at George and Jackie, they followed on the road.

Charlie asked, "Why do you want to take him on in your car. It would be destroyed if it hit that thing."

Steve said, "It's my fight. I'm not worried anyway. If I damage my car I can fix it. If the SUV is damaged it can't fix itself. All I have to do is hit one of the wheels or take out the radiator." He held up a pistol and said, "This is for the radiator."

Charlie asked, "Can't you take out a wheel with it?" Steve laughed and said, "I'll be lucky to get the radiator.

It's not easy to shoot from a moving vehicle." Steve clicked the safety off and held the pistol downward and ready.

Charlie reached behind his seat and pulled out a crossbow. He reached behind his seat again and pulled out a quiver and handed it to Steve. He said, "Hold this." He reached into the quiver and pulled out an arrow. The tip was black, jagged, and metallic. He loaded the quiver with his right hand, bracing it against his knee, while keeping his eyes on the road. The arrow clicked neatly in place. He lifted the crossbow out of the floorboard and held it up. He told Steve, "It's very hard. I'm not going to do it while we are bouncing like this because it takes too long to load and I want my arrows back." Steve nodded with his eyes wide open. Charlie continued driving and held the crossbow with his right hand. Charlie said, "I wish I had a better target. It's too hard to get to the wheels from here and I don't have anything here that would stop him. If I hit his back gate he just gets a free arrow."

Steve answered jokingly, "He probably wouldn't even appreciate it like you do. I don't think he is a collector."

Charlie grinned and said, "Probably not. Let's wait for a better target. Maybe he's feeling generous." The Hummer was now about fifty feet from the back of the SUV. It pulled right hard and passed closely in front of the Hummer. Charlie turned inside the SUV's turn and narrowly missed it.

Steve said, "If you clip him with your front end he will have more damage than you will. Feel free to do that next time. I will gladly pay for your body work." Steve held the pistol out of the window and pointed it at the SUV. He felt his hand bounce around and almost dropped the pistol. He pulled the pistol back inside.

Charlie pulled up alongside the SUV and turned his wheels hard right. He rubbed the left side of the SUV. The SUV applied brakes hard and the Hummer zipped forward and away. Once it was clear of the SUV it started turning hard right and spun around until it was facing the SUV. Charlie stopped his Hummer and both Steve and Charlie popped out of their windows and fired shots into the grill of the SUV. Steve emptied his clip and Charlie launched two arrows. The arrows stuck in place and Steve could see small vapor clouds where his rounds impacted the grill.

Steve shouted, "Yes!"

Charlie asked, "When can I go collect my arrows?" He held his door handle as if he were ready to get out.

Steve yelled, "Not yet!" Steve reached out and held his arm. He explained, "He's not dead. He is crippled. He has to run on batteries only now."

Charlie said, "Shoot him some more."

Steve said, "I can't shoot anymore. I am out of bullets."

Charlie said, "Reload it!"

Steve said, "I'll have to buy some. This isn't my pistol." Charles asked, "Whose is it?"

Steve pointed at the SUV and said in an exasperated yelp, "His?"

Charlie shook his head and looked away, "This is too much. I'll bet there's a really good story there somewhere."

The SUV pulled forward to pass them on Steve's side. Charlie yanked another arrow out of the quiver by Steve's side, reloaded his crossbow and climbed over Steve to get a shot at the tires. Charlie ended up on top of Steve with his knee resting on Steve's thigh and his arms were hanging out of the window. He shot and missed. Steve was frozen in pain because Charlie was resting on a pressure point. The SUV passed and was opening range.

Charlie backed off of Steve and returned to the driver seat. Charlie said, "Let's go!"

Steve was still wincing in pain as he said, "Don't you want your arrow?"

Charlie said, "Yes!" He stopped, jumped out, ran to his arrow, grabbed it, threw it into Steve's window, and returned to the driver seat. He put the Hummer in gear and left a cloud of dirt all around. In a very short time he was fifty feet off of the SUV's quarter panel.

Steve remarked, "Those must be expensive arrows." Charlie said, "Don't ask."

Steve said, "I don't care what it takes. I want to either get those two arrows back for you or replace them. Thank you for the help."

Charlie said, "Thank you for that. I'm just glad to be doing a good thing right now."

Steve chuckled and said, "Those are small words for what you are doing right now. You are helping me stop an incredibly dangerous man before he can do more harm."

Charlie asked, "How bad is this thing?" Steve said, "He used to be a hit man."

Charlie asked another question, "What does this have to do with you? Your plates say Colorado?"

Steve answered, "I killed him and then made him into a set of software. Barron Motor Company turned him into the SUV from heck."

Charlie asked, "Why aren't they out here fixing the problem?

Steve said blankly, "They don't know what to do. They probably hired a building full of experts. Who knows? Maybe those experts will come up with something in a couple of weeks."

Charlie said nonchalantly, "Is there any chance I'll get paid?"

Steve said, "Bad guy first then money! Let's get this guy before he kills anyone else! If you're wondering where the paycheck comes from it's me. If you ram him I'll fix your truck! Does twenty thousand cover the bill?"

Charlie said, "Sure, just wondering. When you mentioned Barron I thought they might help out."

Steve replied, "Only if we're lucky. They might try to pin it on us. If we stop him soon enough, maybe they won't notice." He looked over his shoulder to find the Beast and George's Oldsmobile. They were about half a mile away. The SUV and Hummer were approaching a road.

Larry saw an RV lot to his left. He turned and headed for it. At the far end of the lot was a vehicle trailer. The vehicle trailer had its ramps on the ground and the other end was pointed at a road. The road end of the trailer was raised up on blocks so the jack could be replaced. Larry scanned the entire area since he was still far away enough to decide what to do. The trailer was only about twenty feet from the road, but it was pointed upward because of the ditch it was parked in. The traffic on the road ahead was slow and almost stopped. The slow traffic headed into a mall and the other lane was almost clear. That lane was traffic leaving the mall. The trailer was pointed upward enough that Larry could possibly jump over the slow cars. That looked like a perfect way to escape the Hummer and other two cars. Larry aimed at the trailer and speeded up.

Larry hit the ramps at about sixty miles per hour. His front two wheels buckled as his front end bounced hard upward. The trailer moved forward and the supporting blocks slid. The top support block was now two inches from falling off. The ramp groaned a metallic groan. The top block snapped and popped as it suddenly took a new load. The sound was a little like hitting a piece of cement with a hammer.

Larry's back wheels hit the two ramps and the back end bounced up high. The rear wheels left the trailer and were up in the air. The impact of the rear wheels pushed the trailer further forward. The top supporting block split in two and pieces went everywhere. The front of the trailer dropped down and Larry's front wheels were airborne also. All of

Larry's wheels came down hard and he bounced upward. He still had some upward momentum when he cleared the trailer, but not as much height as he hoped for.

He didn't clear the first car. He landed on its hood with his left front wheel. At sixty miles per hour, the hood was as soft as a pillow, but the engine underneath was not. For the tiniest fraction of a second the other car bore the weight of itself and Larry, then its suspension and wheels failed. Larry didn't land on a flat surface. One wheel went deep into the engine compartment. The other wheel waited to find something to land on because the car wasn't underneath it. Larry began to twist through the air as he bounced off of the engine of the victim car. He twisted in a clockwise corkscrew. He still had some upward momentum and his front lifted higher than his back. Larry panicked and looked all around himself. He saw the ground from the bubble mounted on his roof and thought about how he was going to hit the ground. The turning seemed to be slowing down and Larry was sure that he was going to land on his top and be like an upside down turtle.

Larry felt a shock that shook his entire being. His roof moved to the right and collapsed the entire passenger area. His visual bubble was still intact and he could see a large red shape appear and disappear. The sound of metal to metal collision was harsh and like a cannon firing. The pain of torn wires and ripped solar panels coursed through his being. The force of this new impact twisted Larry harder in a clockwise spiral. All of the new motions added together to land Larry on the road and pointed directly at oncoming traffic. He landed front first, bounced upward, then landed hard on his rear wheels.

A white Ford F-350 Super Duty truck was bearing down on Larry and collision seemed imminent. Larry pushed backward with all of his wheels at full force and turned his front wheels left. He began to move backward and twist toward the shoulder of the road. When the truck hit him, it just pushed him harder into the ditch. The truck stopped and the driver got out. The man began yelling obscenities and said, "Look what you did to my bumper! It's going to cost a fortune to fix that dent!" The other driver approached Larry and looked inside the passenger compartment. He gawked at the mess and said, "What happened to you?" Larry backed away from the man. The man stood in place holding his hat in his hand and scratching his head. As Larry backed away from the driver of the white pickup truck, he saw a large red garbage truck driving away in the distance. Larry thought, "Gee, I could have gone out with the garbage."

Steve and Charlie were very near the vehicle trailer. They were spellbound by the wreck that Larry had and not watching what they were doing. Steve reached over and grabbed Charlie's arm as they were fifty feet from the trailer and said, "Don't do what he did!"

Charlie snapped out of it and pushed his brakes to the floor. The Hummer wasn't going as fast as the SUV was, but its momentum still carried it onto the vehicle trailer. The trailer slid forward again and the blocks collapsed to the ground. The Hummer slid to a stop in a nose down position and rested. The front of the vehicle trailer was all of the way on the ground and the ramps were up in the air. Charlie put the Hummer in reverse and gunned the engine. It launched backward off of the trailer. The bottom of the Hummer scraped the trailer as it left. Charlie said, "Thank goodness for high ground clearance." Steve sighed with relief because visions of high centering on a trailer danced in his head. Steve was glad that he had not tried this with the Beast. Charlie stopped about thirty feet from the trailer and both of them watched the SUV move away. Charlie asked, "Now what?"

Steve pointed at the mall and asked, "Can you get to that without going through that?" He ended by pointing at the congested road.

Charlie answered, "Sure, but it'll cost you." Steve asked, "What?"

Charlie said, "My ticket and my lawyer." Charlie hit his accelerator hard and dirt formed a cloud behind him. He twisted and fishtailed until he was out of the ditch and he picked up speed as he made his way across the brush and bumps. In the distance, Charlie heard sirens and asked Steve, "Think that's for me?"

Steve said, "Probably not. The SUV's wreck was pretty ugly. We might get a little business though because we left tracks close to it."

Charlie reminded Steve, "You are taking care of my lawyer aren't you?"

Steve replied, "As long as we take care of that." The Hummer made its way around the traffic and slid in neatly at the end of the parking lot.

The SUV entered the parking lot at roughly the same time far away. Larry was moving fine, but he noticed that he wasn't charging as well as before. One third of his solar panels were damaged by the garbage truck. Larry moved fairly well through traffic because other drivers stopped when they saw him. They were shocked by his appearance, smell, or maybe just curious. Some drivers actually backed away from him as if they didn't want to get any of him on them.

Larry saw the Hummer at another end of the parking lot and decided to lose it. He wanted to lose them in the mall. He was further from the front door of the mall than the Hummer was, but decided that he could get to it first because he could move like a squad car through the mess of cars. He began to sound his horn furiously and slipped between cars like a hot knife through butter. Cars all around him stopped and looked. Some cars wrecked because they were distracted by the strange sight. A few drivers rolled their windows up as Larry got close to them. Flying insects left the SUV and attacked other cars. A couple of mice jumped out and scurried under parked cars. The squirrel which had been hiding in the back while Steve was inside jumped out into an open window of another car. The driver of the other vehicle stopped suddenly. He twitched about and lifted up out of his seat. He and his passengers screamed loudly as the squirrel moved around the inside of the vehicle. All of the doors of the other vehicle opened and everyone jumped out. A vehicle which had been following the SUV stopped suddenly to avoid hitting one of the passengers jumping out of the squirrel infected car.

Steve and Charlie approached the entrance of the mall. Steve pointed and shouted, "He is over there! He is going to go inside the mall. Can we stop him?" Charlie pulled the Hummer onto the sidewalk and tried to move toward the door. He leaned on his horn to get people to get out of his way. People all around turned and jumped out of the way. They still didn't move fast enough for the Hummer to get to the mall door first.

Larry flew through the front doors of the mall. The glass shattered and steel door frames bent like spaghetti. One of the pieces of steel frame wrapped around the right rear wheel of the SUV and dragged on the floor. It ripped tiles off of the floor as the SUV moved into the mall. People all around Larry scattered in panic. A woman with a baby stroller and infant froze in place, not knowing what to do. She turned around to run away and tripped. The baby stroller ran over her ankle. She got up and untangled from the device, then she ran away pulling the stroller with one hand. Her shopping bags which were hanging low to the ground hit her in the legs and tripped her every couple of steps as she made her way away from the SUV. She reached the wall and leaned against it as if it were a safe dock in a violent storm. She turned around to see the SUV. It was far enough away that it wasn't going to hit her. She watched it pass by. As soon as she felt safe, she began to move along the wall to try to get out of the building. She was shaking like a leaf and looking over her shoulder as she fled.

Steve and Charlie sat outside of the mall for a moment. Charlie said, "What now? If we follow him, they will blame us for all of this. If he gets away, then we will be holding the bag. I have plates. He doesn't. Do we really want to do this?"

Steve said, "We don't want to do this!" He looked confused and exclaimed in disbelief, "I don't know what to do! What do I do? I can't go in there! I can't let him go! I have to stop him! I can't stop him! What will I do if I meet him inside? Why are you asking me?" He paused for a moment and said, "We have to go inside. We have to follow! We can't just let him go. It doesn't matter if they blame us. People are going to die if we don't!" He grabbed Charlie by the arm and said, "Let me drive! It will be my fault! Don't ask, just do it! Now get out!"

Charlie stopped the engine and pulled the keys out of the ignition. He opened the door and leaped out. He ran around to the other side of the Hummer and got in. He handed Steve the keys and said, "You didn't think I'd let you go anywhere without me did you?"

Steve replied calmly, "Of course not. I would never think such a thing."

Charlie reminded Steve, "You are still going to pay for my lawyer aren't you?"

Steve laughed as he spoke, "Grow some courage will you? Life means taking chances."

Charlie yelled, "Courage? What did I just do? Is anyone else here doing anything better? We could step outside and discuss this if you'd like."

Steve laughed and said, "I do like. I like you better already." He pinched Charlie on the left cheek and said,

"You are beautiful!" Steve started the Hummer and left a few pounds of rubber at the wide opening to the mall left by the SUV. He said, "Oops! This is very nice and I like the power. I think I want to have one of these. I'm definitely falling in love with this vehicle."

Charlie yelled, "Just shut up and watch where you're going!" He was upset because he was not in the driver seat.

Steve and Charlie passed the lady with the baby stroller and glanced at her. They moved to the opposite side to give her room. She stopped and stared at them for a moment. She waited for them to pass before moving again. This time she ran out of the door screaming. Steve shouted excitedly to Charlie, "Hey tracks! He left tracks!" They both looked at the trail of ripped tiles left by the piece of the entrance door wrapped around the SUV's rear wheel.

Charlie shouted, "Good for us! You can follow them slowly without running over anybody! Slow down and take your time. We can catch up once we are out in the open again!" He paused for a moment and shouted, "Now!"

Steve listened and slowed down. He said calmly, "Yes sir." Not only was this Charlie's vehicle, but he was right.

The trail of ripped tiles went to the right and through a jewelry store. Two men were in front of the store yelling and pacing back and forth. Glass and pieces of jewelry littered the floor. A column which was at the corner of the jewelry store was in pieces and mixed in with the other debris. The two men who had been pacing in front of the door saw the Hummer and moved out of the way. They shouted, "Maniac" as Steve and Charlie passed by.

On the left side of the Hummer was a guard rail and open space where they could see the level below. Charlie said as he looked down to the lower level, "I hope this floor can handle the load."

This corner was narrow and despite the missing column, there was very little room without hitting something. The Hummer actually scraped the guard rail before Steve turned right to avoid going over the edge. Steve grumbled, "Why did I come in here?"

Charlie answered dryly, "Because there is a great sale at Sports-R-Us sports gear."

Steve said, "Yes! I want to buy a dozen arrows for your crossbow."

Charlie agreed, "Yes, that's right. That's a good idea. Maybe we could find some ammo for your pistol. I mean his pistol."

Steve said, "Maybe we could." Steve and Charlie continued to follow the trail around the corner.

Larry was in a large department store. He knocked over shelves and racks of clothes as he made his way through the store. He pulled up to a service counter. A woman behind the counter watched him pull toward her. She was confused and gave an automatic response, "May I help you Madam?"

Larry shouted, "Yes! Change the last part to Sir!"

She trembled as she spoke, "Yes sir. May I help you?" Larry spoke softly but clearly, "Yes you may. Get an extension cord, plug it in, and then plug the other end into me. If you do what I say, then you will be able to work here tomorrow because I won't drive through this counter and cause you to be moved to restocking shelves in lingerie for the next two months."

The woman thought briefly about his proposition. She didn't like standing behind the counter but restocking was a lot worse. Larry was lying because he didn't know what the store would do to her, but she believed his threat. She agreed quickly and said, "Yes sir." She rushed to hardware. She wanted to tell a manager, but didn't have to. Store employees from all departments rushed to the area and stood a respectable distance from the SUV. In minutes a thirty foot wide circle of people formed around the SUV. She plugged the end of the extension cord into the socket at her station and opened the extension cord up to full length. She walked the other end of the cord out of her station and approached Larry. Larry opened a small door next to the opening for terminal 205. A set of sockets was inside. The top socket was female and the bottom one was male. The label above them said, "Charge 115 Volt."

Larry felt the surge of electricity flow into his charging system. He felt good all over. Despite his injuries, he felt euphoric. He said, "Oh, yes. I like this. Thank you." One of the store employees in the circle around Larry moved a little closer. Larry scolded the employee, "Naughty, naughty. Don't do that. If you behave then I will. I will leave nicely and not hurt anyone. If you can live with that then so can I."

The manager of the store reached out, grabbed the arm of the person who stepped forward, then gently pulled that person back. The daring employee backed up reluctantly.

Larry said, "The last person who got in was really anxious to get out. He couldn't get out because I was going to run over him if I had a chance. If you do get in, then you'd better bring along a lot of air fresheners." The person who had stepped forward before stepped back behind the circle. A mouse climbed over the top of the dashboard and onto the hood. It scurried to the end of the hood and jumped onto the counter in front of Larry. It disappeared into one of the displays. The lady behind the counter backed away from the mouse and cringed, but didn't go very far because she was afraid of Larry.

Five minutes after Larry arrived, he said, "I feel so much better now. I will go nicely and leave you all alone. Just give me room and I'll back out of here." He began to back away from the counter. Larry had only charged a little bit, but replaced some of what he used in the mall. The cord straightened out, began to stretch, and then popped out of the wall. The female end of the cord remained attached to Larry and he had a long tail dangling until he ran over it and it popped out too. Larry backed out the same way he came. He passed racks and shelves and used his rear pointing cameras to find his way out. Larry listened to the police broadcasts while he was charging and he knew that the upper level of the department store was surrounded. He heard them say that he couldn't go down to the lower level because he couldn't fit in the elevator. Larry saw the Hummer approach from the other direction as he was leaving. They were coming through the main door of the department store.

Steve and Charlie noticed the SUV right away because they were looking for it. Steve pointed at the SUV trying to hit it and gunned the accelerator hard. Larry saw what was happening. He turned his wheels sharply left and aimed toward a part of the wall where he figured there wouldn't be any steel girders. He broke through the wall cleanly with the rear bumper and door. The sound of the wall breaking was like a rockslide. Pieces of gypsum board and brick veneer littered the area in front of the store. The Hummer missed the SUV and stopped neatly without hitting anything. The top of the Hummer stopped just short of the front door and did not hit the top steel supports.

Charlie leaned over and said, "Thank you for not destroying my windshield." Larry stormed down the aisle away from the department store. The piece of door which had wrapped around the rear wheel didn't tear up tile while the SUV was in reverse. Charlie watched the SUV make its escape and yelled, "What are you waiting for? Come on. Let's go get him."

Steve began to follow the SUV in reverse but didn't do very well. He approached a narrow spot and stopped before hitting a guard rail and falling to the lower level. He pulled forward and turned around. He then followed as fast as he could without hitting anything. The SUV drove better in reverse than the Hummer because with its rear view cameras it could see better than Steve who had to turn around to see where he was going. The SUV was also about a foot narrower than the Hummer.

Larry saw a police cruiser trying to cut him off. It stopped at the end of the narrow aisle Larry had to get through to escape. Behind Larry was the Hummer and closing fast. In front of Larry was a squad car sitting with the passenger door of the vehicle blocking his way. Both policemen left the squad car and stood on the driver side of it because that was standard procedure. The theory behind this move was that the car would protect them from anything creative that the suspect would do. Both policemen stood with their weapons ready. Their target of choice was low on the tail and hopefully the gas tank. They couldn't hit a driver and the wheels were not available as a good shot yet because there was too much interference. They shouted out, "Stop Police!"

They just finished the sentence before Larry speeded up and slammed into the passenger door of the car. The car bounced suddenly toward the two policemen. The policeman who was leaning against the car was thrown forcefully backward. The man broke numerous bones and injured his neck and back as he landed on the floor. He slid several feet before he hit another guard rail leading to the lower level. The policeman who was about a foot away from the car was able to jump back before the car hit him and he did not have any serious injuries.

Larry's push was just enough to spin the police car around and out of the way. It was now to his left as he faced forward. The engine of the squad car was heavier than the back of the car and it didn't move as fast as the back. The squad car rested against a wall of the mall. The policeman who was off balance and not hurt very badly fired several shots at Larry. He didn't hit the wheels. He did hit the gasoline tank and gasoline began to leak on the floor. Larry knew that it had been hit, but didn't worry about it. He couldn't do much with it anyway.

Another squad car approached Larry as he came out of the narrow passage. It was to his right. He aimed directly for it hoping to ruin its radiator with his tail end. As he turned to hit the other squad car, he scraped the squad car against the wall with his front. Once clear of the empty squad car against the wall, he twisted quickly and slammed hard into the approaching squad car. He pushed that squad car against the mall wall on the other side. The officer in the second squad car had his arm out of the passenger window and it was crushed. He couldn't move at all. The driver of the car was fine because he had a seatbelt on and the front of the vehicle was hit. He fought with the door because the fender had been pushed against it. He finally got free. He rolled out of the driver side window and collected himself. Larry saw an opening and moved forward through the next tight spot of railing. The two officers who were not injured fired shots at Larry but only hit his tailgate. The Hummer closed in on Larry, but missed, and Larry managed to get into the next tight passage. Larry now passed through this tight area going forward. The Hummer was now approaching the tight area in pursuit of the SUV.

The two uninjured officers ran to the Hummer and yelled, "Stop!" Steve stopped. The two officers opened the rear driver side door and jumped in.

The more experienced of the two officers asked Steve, "Why are you here?"

Steve answered, "I want to stop that!"

The officer answered, "Yeah, me too! Let's go!"

As Larry left the tight spot, another police car approached him from his left. Larry spun to the left and drove the car into the wall hard. He then backed hard and pushed the Hummer into the other wall. Ahead was a set of stairs. They were wide and strong. They were steep too. Larry made up his mind. He decided to try to go as slowly as possible down the stairs. He didn't want to go in reverse because he could end up stuck at the bottom with his rear bumper pushed way forward. He knew that the police thoroughly covered the top level of the mall, but had only a couple of cars below. He tried his luck. He speeded up to ten miles per hour and went to the stairs.

His front wheels fell off of the end of the stairs. He hit his frame hard. He could feel sensors screaming as he did this. He continued to roll forward because he still had momentum. He also pushed as hard as he could with his rear wheels, because he didn't want to be stuck there where the policemen could destroy him. After all, he wouldn't even get a trial. By the time his rear wheels could not touch the ground any more gravity took over. The front of the vehicle was heavier and it tipped slowly forward. His front wheels bounced hard and then his rear wheels hit the stairs. It was too late to go back and he didn't have much control anymore. He hoped that he wouldn't go over the side of the set of stairs. He bounced clumsily like a slinky as he descended. His sensors screamed and he actually felt motion sick since confused signals went to his brain from what was happening. He rolled from side to side and the hand rails saved him a couple of times as he sunk lower and lower. His brakes didn't work and he panicked for a moment when he realized he had no control. He thought, "I am almost there! I am almost there!"

Steve, Charlie and the two policemen got out of the Hummer and went to the stairs. Steve followed the SUV down several steps and bent down very low to try to shoot one of the tires. His pistol went, "Click." The two officers fired at the tailgate and Steve grabbed one of them by the collar. He pushed with anger and full force until he had the officer against one of the mall walls. He spoke softly, but deliberately, "He is running on batteries. Only a shot to a tire would slow him down!" Steve released the officer and looked to his left. The officer's pistol was aimed at his head. Steve said in frustration, "Do you really think that would help anything?" The officer lowered his weapon and relaxed. The other officer followed Larry down the stairs with his pistol at the ready position. Steve noticed what was happening and ran down the stairs to follow the policeman. He yelled, "Get back up the stairs! He will run over you. Come back now!" The officer looked back with a confused look on his face and turned around to continue chase. Steve yelled again, "You don't have a car to protect you. He will kill you!" the officer ignored Steve. He jumped and landed on top of the SUV. Steve yelled again, "Get off of him while you still can. He will kill you at the first chance!" The policeman didn't have a very firm grip because there was very little to hold on to and the bouncing was very rough. The officer bounced up and down as violently as a basketball. He bounced forward over the roof and hood of the SUV. He finally fell and landed on the stairs below the SUV. Larry bounced hard on top of him at the bottom of the stairs. The policeman landed on his right side on the floor below. Larry fell on top of him with his bumper. He crushed the man's rib cage and stopped his heart. When his wheels hit him, they rolled over him pushing his head and feet off. Blood splattered in all directions and covered Larry's fenders. The bloody mess did not move very long. The man didn't even have time to scream as he died.

Larry's front wheels were now firmly on the lower level. Larry pulled himself off of the last few stairs with his front wheels because his rear wheels were not on the ground. He moved slowly at first, but managed to go down another step.

Finally his rear wheels barely touched the floor and he turned them to pull his bumper off of the stairs. He was free. His bumper bounced hard on the last step and he rolled away. The last step cracked in several places from his bumper and most of it crumbled to the floor.

The entire area which, moments ago had been full of people was suddenly empty. The death and blood was too much for them. Only a few people were nearby at all. A mall security guard stood and watched with his mouth open and not knowing what to do. He held his radio up to his face and keyed the mike, but said nothing. Two people who were in front of Larry backed cautiously away. A woman holding a small child sat on a bench a few feet away from Larry. The child struggled to get away, but she held him close and whispered in his ear to be quiet. Blood from the dead policeman covered her dress and shoes. The child had blood on his pants.

Larry rolled smoothly and quietly away. He headed for a large aisle because he assumed that there was an exit there. A woman nearby picked up a few chunks from the broken bottom step of the stairs and hurled them at Larry. Larry kept on going and speeded up. She screamed and ran after Larry. She caught up with Larry and beat against his quarter panel with her fists. She had nothing to hold onto and slipped. She was on the right side of Larry and grabbed anything she could reach. She pulled the door open where the plugs were for charging. The small door came off in her hand. She cut her hand as she did this. Her blood covered the side of Larry. She hit Larry several more times and left handprints in blood near that door. She grabbed another door on Larry's side. Inside was a round white object. At the bottom were small long things like legs. The door was stronger than the other one. She held on to keep up with Larry because he was going faster. Her feet slid and she was almost water skiing on the smooth tiled floor. A foot away from her the tiles were coming up because of the piece of the other door which was still wrapped around the back wheel. She held on tight to the door and it finally gave way. She fell forward and landed on her right shoulder. She stayed in the ground rolling in pain and holding the door in her hands. Larry's wheel missed her by inches.

On the level above Steve and Charlie looked at each other and wondered what to do. Steve yelled out, "You! Come with us!" He pointed to the policeman who was still with him. The policeman pointed at himself with a question mark on his face. Steve ran up to him and grabbed him by the collar. He said slowly and calmly to be understood, "You have to get us out of here. We don't fit on that. Your vehicle is down. You are commandeering us. Now move it!" He pushed the officer toward the back door of the Hummer, opened it, and gave a final push to let the man know where he wanted him. The officer didn't know how to respond to this so he went along with it. The officer believed Steve because he had seen that Steve was right about the other officer and in this situation seemed to know what he was doing. Steve said, "You are coming along for the ride. You need to tell your buddies to let us out of here so we can chase the bad guy. Right now there isn't time to be nice." Steve slammed the door on his new guest and got in. He saw Charlie and realized Charlie couldn't get in because the Hummer was against a wall. He got out and motioned Charlie inside. Charlie got in, then Steve, and then the Hummer sprang to life again. By the time the Hummer was five feet from the wall everyone had seatbelts on and held on tight. Steve reached the exit in no time. He stopped for a uniformed policeman. His new passenger gave a quick explanation and they were off to reach the lower level of the mall. The policeman in the back seat spoke up, "Who are you and what is going on? Where is the driver of that vehicle?

Why are you chasing it?"

Steve gave short answers because he was maneuvering to get to the lower level and away from traffic, "I am the man who made a software set out of a man. His entire mind is on disc. That set of discs became a car. The man was a contract killer. Would you like for me to continue chasing it?"

The policeman paused for a moment while he absorbed what Steve said, Then he said, "Yes, please!"

Steve was in the middle of a turn and didn't answer right away. In a moment, he said, "Thank you. I will." At the end of the turn he saw the SUV leaving the parking lot and going out into a field. Rain drops appeared on the windshield and Steve turned on the windshield wipers and lights. A moment after he did this he shouted out, "Yes! It is raining! It doesn't get any better than this. He is toast!"

CHAPTER 17

JUNKYARD

Charlie was confused and asked, "Why is this special?"
Steve answered, "He doesn't have a windshield and his
wiring is in big trouble. If we get enough rain it will kill
him." Steve paused and answered a little better, "He wouldn't
have as much control over the vehicle. He will be unable to
maneuver and escape." Steve was still very far from Larry
and fighting through the parking lot to get to him. Other
cars got in the way and slowed him down.

Larry knew what was happening and was scared.
Water drops blurred his vision on the dome on his roof and
he had to spin it to clear it. The ground was still firm, but
as it got wetter it was getting softer and his tracks were more
visible. He found a road at the end of the next field. He
got on it and went full speed to the right. He didn't care
where North was or anything else. He needed to get away
from that Hummer. He left a few tracks for half a mile until
he slung the mud off of his tires. He kept going to take
advantage of his lead. He scanned maps of the area and tried

to guess where he was, but he was more concerned about finding shelter from the rain.

His luck held up. There was a junkyard to his right. He backed up and went down the dirt road to the gate. Everyone was distracted by the weather and getting inside. Normally a vehicle like Larry moving under its own power would have attracted a lot of attention in a place like this. By all rights, he should have been towed in, but everyone was running inside to hide from the rain. Most of the vehicles around him were taller than him and very dirty so he was able to hide as he carefully slid into the opening of the yard. Larry was quiet too. His wheels only made a muddy sound and he didn't use an engine at all. It was as if he were dragged in. Larry managed to get behind the first row of cars without being noticed. Nobody was beyond that point because they were inside waiting for the rain to stop. The rain was loud too. The raindrops pounded the metal roof of the shop where everyone was. Nobody heard enough to pay attention.

Larry shopped for a hiding place. He went through several rows of cars before he found the perfect spot. A stack of cars had a large gap underneath. It was against the back fence. He slid in perfectly. Rain still poured in through a few places. It was bad, but acceptable. Several relays were slow to respond now. Larry turned them off and hoped that they would dry out and work again later. His right side where the bloody handprints were was against the fence, so it wasn't obvious. The policeman's blood on his fender was buried behind several cars. Larry listened to the rain and waited. He worried about his pursuers and the damage from the rain.

As Larry sat there a puddle formed in the rear seat well. The dirt under his rear wheels became soft and shifted from the rain. His rear tires sank about an inch into the mud. After fifteen minutes of sitting there, he began to blend into the surroundings so well, that anyone would have thought that he had been there for years. The white globe on his right side, which was now exposed because the woman pulled the door off, stared at the fence. The little cameras moved around to see. One of the legs moved. It slid forward very slightly and then back again to safety.

The only stimulus available was the rain at the moment. The animals and insects inside Larry noticed that he wasn't moving anymore but the rain was coming in. He heard tiny footsteps inside. A bird inside fluttered its wings as it found a better place to nest. It moved randomly when it found and attacked an insect. Mice moved back and forth. Some of them were eating bugs and there was a piece of the first driver under the driver seat. A few more animals came in from other cars in the stack. A stray cat crawled across the hood of the SUV and dropped in through what used to be the windshield. It leaped and ricocheted off of seats as it chased mice and birds. Chewing, biting and clawing creatures ripped up the upholstery and the blood stained felt lining for the roof fell to drape over the seats. Cockroaches fell from this opening and other animals chased them for food. The mildew odor and the smell of all of the animals were beginning to cover the smell of the first driver.

The white globe which was Larry's brain heard all of these sounds, but what upset him the most were the bugs in his own little compartment. A spider began to weave a web across the little cameras mounted on the globe. It liked the location because it was out of the rain. The spider snagged a cockroach and draped its corpse across one of the cameras. Larry waved one of his little arms furiously but couldn't reach the intruder. He lifted the entire globe to shake it loose, but couldn't do it. Larry was stuck with an unpleasant view in his right eye.

Steve, Charlie, and the policeman drove out of the parking lot and followed as far as they could until the trail ran cold on the other road. They looked for tire marks on both sides of the road for several miles. They went up and down the same road three times before they turned around to go back to the mall. George and Jackie were at the mall parking lot near the spot where the SUV and Hummer left it. There was a clear path where both vehicles tore up the landscaping and proceeded through the field. George and Jackie were in their cars waiting for the rain to stop. When the Hummer returned everyone piled into it and discussed what to do.

The policeman in the back seat of the Hummer scooted over for George and Jackie. He asked Steve, "What can you tell me about this thing we followed?"

Steve answered, "I made a set of software. I gave it to one of your senior people. It contains the whole life of a hit man. Barron used the software to build a vehicle which would drive itself. The personality on the discs took that chance to build his entire brain into the vehicle so he could live again. The Barron people had no idea what was going on until it was too late. The vehicle runs on a large bank of batteries so it is very quiet. It has a gasoline engine, another engine and solar cells to charge the batteries. The radiator is ruined so it runs on the solar cells only. It needs a long time to charge now because some of the cells are ruined too. It wants to hide, charge, and escape. Does that answer your questions?"

The policeman asked, "What other engine?" Steve looked at George and said, "George?"

George said, "Propane. It won't do anything now if the radiator is destroyed."

Charlie held up his crossbow and said, "It is ruined all right. That thing owes me two arrows."

The officer asked, "What about the software set you talked about?"

Steve answered, "Go find out from your people. While you are at it, I want it back. While I have you here, give me a way to reach you so I can find out what your people are doing about this."

The officer pulled a card out of his pocket and handed it to Steve. He said, "You need to stick around to answer questions."

Steve said, "I'll be here until that thing is destroyed. If you have any more questions you can ask Barron. It is their property. They are really the ones liable for what it is doing. I didn't tell them to bring the guy to life."

George spoke up, "I did that."

The policeman asked Steve, "Why are you here then?"

Steve said, "He came after me and I killed him. It was a job before. My ex wife hired him to kill me. I got him first. I have to stop him now before he comes after me again."

The policeman said, "I don't have any more questions right now. Where are you staying?"

George handed him a card from the hotel they were staying at and one of his own cards. George said, "I think we are all going back to the hotel now. I can't imagine that we could do anything else anyway. We are kind of rained out." The policeman opened the door of the Hummer and left to rejoin the rest of the police at the mall.

Steve looked around at everyone else and said, "I guess George is right. Let's meet at my room and decide what to do." Everyone returned to their vehicles and drove back to the hotel. On the way, Steve stopped at a bank to pull out money for the damage to Charlie's Hummer, payment for Charlie's help, and some extra cash to use while he was there.

Charlie took the money and thanked Steve, "Thank you for the money. I expected to get it from Barron though. Does this mean I am finished?"

Steve said, "Only if you want to be. We could still use your help though."

Charlie nodded and said, "Sounds good to me." They all proceeded to the hotel and stayed there talking.

CHAPTER 18

THE PET

In another part of this town, Theresa Mills got home from the hospital. When she got into the front door of her apartment she threw her purse down on the table.

She was in her living room as soon as she walked in the door. The large computer on several desks in the living room was on. The computer was her living room. A smaller computer was in a corner of the living room near the kitchen. External CD drives and hard drives filled shelves everywhere. Software discs filled more shelves. A small shelf was full of woman's choice movies.

Theresa or Terry was single and 49 years old. She looked a little bit like Kobe Bryant with very large glasses. Her figure was slight. It was so slight that if she didn't wear a dress she sometimes was confused with a man. She loved her work and didn't like going out. This was a habit she started when she was younger and couldn't find dates. She wanted to have someone in her life, but never found "Prince Charming". She never found "Anyone Charming". She would have settled for someone considerate, but nobody ever was to her.

On top of a stack of books was a prototype of the dome on the SUV's roof. The cameras moved and looked at her. A voice from many computer speakers throughout the room spoke to her, "Welcome back Sweet Heart. Where have you been?" The sound paused for a moment and asked, "What happened to you?"

She yelled, "You did you jerk! You burned the place down and left me to die! I have a good mind to fill the RAM with Bernie the Bug videos."

The voice on the speakers answered, "You don't have any. I would know. I've been staring at them for the last few days." He thought about how many times he tolerated those videos from friends and how he would have busted a cap in Bernie for free.

She growled, "I'll get some!"

The voice paused for a moment and thought. He knew that she could push the off switch and he would be dead. He decided to turn her anger down a few notches. He answered softly, "I did no such thing. I have been right here waiting for you while you were gone. I had nobody to talk to and finally the only thing I had to do was watch your fish die. I miss them so."

She walked over to the fish bowl and moaned, "Einstein and Edison!"

She marched back to the dome where the cameras were and said, "I would have been here to feed them if I hadn't been in that fire at the company."

The computer responded, "Fire? What happened?"

She explained, "They brought a driver in to test the vehicle and it went nuts."

The computer said, "I wasn't there, but I can guess that the computer and vehicle felt threatened. I'm just guessing here, but if something happened that would kill you, would you defend yourself?"

She replied, "Yes."

He asked, "If you knew how to kill would you?" She said, "Yes I would if it saved my life."

He asked further, "Did anything happen which would have indicated a threat? Did the vehicle give any warnings?"

She said, "Yes, it told the driver to get out." He asked, "Did he?"

Dead Memories

She replied, "No."

The computer responded, "Then the driver was a threat and would not leave. The vehicle defended itself. If the driver had left, would there have been a fire?"

She said, "I don't think so."

The computer replied, "Then it tried to stay alive and the driver resisted. You would defend yourself too. If a computer was going to kill you, wouldn't you destroy it?"

She said, "Yes I would."

The computer said, "Then why are you mad at me for something you would have done. I wasn't even there and you came in mad at me. I have been sitting on this set of desks waiting for you. I even made something for you." He paused for a moment while he accessed the paint program and drew a rose. He then displayed it for her and said, "Look at the screen and see what I made for you."

She looked at the screen and wasn't angry anymore. She appreciated the gesture and knew that it was all that he could do. She said, "Aw."

He said, "Don't you see that I care?"

She replied, "Yes I do. I'm sorry that I got mad at you. I know you can't do anything here on my desk. Let's make up. I'll put in a good movie for us." She walked over to the small shelf and pulled out, "The Truth about Cats and Dogs." She put it in the DVD player and went over to the stack of books. She lifted the dome prototype up and carried it over to the couch. She sat down on the couch and placed it next to her.

He interrupted, "Before you start this, why did your purse look bigger?"

She stood up and walked over to the table where she threw her purse and pulled out a CD carrier. It was full. She said, "It is you. I saved the software set from the fire. Everything you are is on these discs. I'll bet I could get a lot of money for these."

Larry silently thought, "I'd give you $100,000 to burn them."

She said, "I don't know if I could ever part with you." She held the globe close and hugged it.

Larry thought, "I could part with you if I could get you to plug in the Ethernet cable. I would virus your computer on the way out too."

She started the movie with her remote and sat back on the couch. She held the globe on her lap like a cat.

Larry sat there helplessly watching the movie. He thought, "The truth about dorks and knuckleheads. Lady, the off switch is over there. Help! Please! I am sick of being your pet!"

CHAPTER 19

CHASE

The other Larry sat at the junkyard thinking about his situation. It was getting darker and the rain became lighter. He heard sounds in the yard which indicated to him that the junkyard was closing for the day. The junkyard dogs were now released to guard the yard. Both of them went straight to Larry because they heard him come in and they heard the sounds of all of the animals inside. The dominant of the two dogs approached Larry and sniffed at him. The dog circled him and took an interest in the bloody handprints on the other side. He licked some of the blood off and wagged his tail. The other dog investigated Larry too. The second dog found the white globe tucked into Larry's side and pawed at it. The dog tried to bite the globe but couldn't get his mouth around it. He heard the sounds of things happening inside and wanted to pull it out to investigate it. In this yard, nothing electronic worked, so this was very curious indeed.

Larry's white globe watched the dog attack it. The dog knocked the cockroach off of the camera. Now the tiny corpse dangled from the side of the camera. The spider retreated deep inside the compartment. The cameras on the globe watched the dog lunge at it and the paws push against it. Larry kept the globe perfectly still and tried not to think very hard. He didn't dare move any of the legs because he knew that the dog would yank the globe out of the hole by a leg and possibly disconnect it from the SUV. If that happened, then Larry wouldn't be able to get the globe back inside the vehicle. He would be disconnected from the SUV and completely helpless. Once removed from the SUV, the globe only had enough battery charge to last a few days. Larry knew that the dog could kill him.

The dog lost interest in the globe and went to the front of the SUV. There was only a clearance of six inches between the hood of Larry and the underside of the vehicle above it. The dogs couldn't get into this space, but here was an opening around Larry on the sides and enough ground clearance underneath to lie down and rest out of the rain. Both dogs found comfortable spots there to rest.

The cat inside moved and the dogs became alert. The cat was full and wanted to find a better spot to rest in. he didn't like being in the yard and so close to dogs, so he slipped out of the spot where the windshield had been. He moved slowly along the hood and climbed up into a hole in the floorboard of the vehicle above. The dogs went crazy. They stood up on their hind legs and clawed at the fenders trying to get firm footing to climb into the space where the cat was. Their heads were in this area and they lunged at the cat. The larger of the two dogs moved to the front of the SUV and clawed at the hood and grill. One of the two arrows fell to the ground as he did this. The dogs ripped white paint off where they scratched at the vehicle. White paint chips fell to the ground. The dogs back paws pushed those chips into the dry dirt there as they danced back and forth trying to reach the cat. The dogs growled and barked furiously.

The cat escaped upward into the vehicle above and vaulted up onto the hood of the vehicle it was in. it was now about eight feet above the ground and looking at the ground below. It hissed at the dogs. Both dogs came out from underneath the vehicle to find the cat. They looked up and saw it. The dogs pawed furiously at the nearest vehicle trying to climb to the cat. They howled and growled in frustration. The cat was enjoying his moment. He teased the dogs from his new spot. He walked along the edge of the hood so the dogs could see him and leaped over the fence into a nearby tree. He slowly made his way down the tree to the ground.

The dogs moved under the stack of cars to the other side to watch the cat. They saw it on the other side of the fence. The cat hissed at them to tease them and turned to walk away. It held its tail high and wagged it slowly back and forth as it triumphantly paraded away from them. The dogs growled and yelped as they pawed and bit at the fence separating them from lunch.

Larry watched the scene and quietly laughed at the dogs losing the battle. His thinking made a little bit of noise.

The dog which had tried to reach him before returned and lunged at him again. He charged the ball a few more times and finally gave up. Larry was relieved to be left alone again and sat there quietly waiting for morning. The rain stopped a few hours later. The rest of the night was quiet and Larry was especially quiet so as not to disturb the two dogs sleeping under him.

Morning arrived and when the owner returned to the yard he called his dogs back to their kennel. Business during the work week was fairly slow so there were not very many visitors to this part of the yard. Local repair shops called for parts and late in the day some people with car problems looked around for parts. Hardly anyone explored the yard looking for projects during the week. Larry sat there in his perfect hiding place in peace and charged his batteries.

Steve, Jackie, George, and Charlie looked all over the countryside to find hiding places with their satellite maps. After two days Charlie told Steve, "I can't keep doing this anymore. Call me if you see this thing."

Steve was disappointed but agreed, "You're right. We have exhausted just about every hiding place. If he hasn't moved much, then he is fully charged by now. I don't know if I can keep this up anymore. I'm going to go back to the hotel and rest for a while." The convoy stopped by the side of the road; everyone got out of their vehicles to talk, and then turned around.

On Saturday business at the junkyard picked up. People walked up and down the rows of cars. They opened doors, lifted hoods, and bent down to look under cars.

They were busy with tools and climbing into the rusted hulks. Larry's row was no exception. Two young men busied themselves with an old Chevy truck. They turned wrenches and tugged at the transmission. One of them said, "It is coming!"

The other young man said, "Just pull! It will land in the soft dirt. I don't need the pan anyway."

The other guy moved clear so it wouldn't fall on him and gave it a good yank. His legs twisted in the air and he kicked with his left leg to yank harder. The big piece of metal crashed to the ground with a thump like a large boulder. The pan on the transmission cracked and a red puddle formed on the ground. The puddle quickly moved toward the young man under the truck and soaked into his shirt before he could move clear. He did move, but too late. He was covered in transmission fluid. He got out from under the truck and told his partner, "You get to wear the cherry juice next time." His friend said, "Nah! It'll be oil next time. I saw a sweet 454 a few rows down."

The soaking wet guy said, "Turbo 400, 454, and a full ton rear end. I feel like accessorizing. Do you think we'll find a fuel injection system?"

His partner said, "While you are feeling sensitive, maybe we should get some better seats for that abortion of yours."

The guy who had been under the truck said, "How about a little respect Carl? You've been getting to work all week in Brenda. She is better than your truck which is up on blocks right now."

Carl said, "I'm sorry that I insulted your sweetheart Barry. I appreciate the ride. Since I'm chipping in on the parts I would call that respect." He turned around and looked at Larry. He stopped what he was doing and walked nearer to it. He pointed at Larry and said, "What is that?"

Barry looked and said, "That's a new." He looked harder and said, "Maybe it's a Nissan."

Carl walked around the back and said, "I don't know. Have you ever heard of a Broadsword?" Barry followed and looked at the back. Carl walked around to the other side. He looked at the bloody handprints and said, "What happened here?" He turned to the compartment with the brain and said, "Hey check this out!"

Larry listened to what was going on and watched as they approached his brain. He thought as quietly as he could and just hoped that they would go away.

Barry looked at the round white ball and said, "I don't know what that is. It looks like some kind of space gizmo. Look at the leg looking things on the bottom."

Carl reached inside the compartment and put both hands around the ball. The legs pushed hard into the sides of the compartment and buried the tips into the metal walls. Carl pulled harder to try to get it out. He said, "This thing is making noise and the legs moved. It doesn't want to come out. Barry said, "I'll get a crow bar."

Larry said through the stereo, "Let the ball go and go away. Take your transmission and don't come back. If you leave now I won't hurt you."

Barry said, "Did the truck talk to us?"

Carl let the ball go and walked around to the front to see inside the passenger compartment. He couldn't get his head into the narrow space, but he could see a little bit. He spoke into the compartment, "Anybody in there?"

Barry moved closer to Carl to see what was inside. Both of them were crouched in the small space between the vehicles. Barry said, "Who said that?"

Larry warned them again, "Leave now!"

Carl reached his hand into the small area where he couldn't fit his head. He put his hand inside the passenger compartment and felt around. He said, "I'll bet there's something in here."

Larry turned his wheels to the left and backed at full speed. The right fender crushed the two men who were exploring Larry. Carl, who was in front, had his left thigh yanked off by the right tire. As Carl fell down the tire went through his torso and finally crushed his head as it pulled him to the ground. The only sound was the spinning of the tire against what sounded like mud. Barry was caught between Larry and the next car. He was crushed by Larry's weight and force. He tried to scream, but just made a choking sound. Once the tire was past Carl, it began to grind its way through Barry turning him to hamburger meat. Barry was unable to move very much, but was still alive enough to make gurgling sounds as his body was torn to shreds. Barry's left arm hit the top of Larry's fender to get him to stop, but Larry kept going. Barry's arm stopped once there was no life left in him. Blood from both men covered the entire side of Larry from the front door forward. The rim was red with blood and the treads of the right tire were full of guts.

Larry twisted out of the opening with such force that he bounced off of cars on the other side. He backed a few more feet to clear them and pushed forward with full force. Without Larry against them anymore, the two dead bodies in the hole under the cars slumped to the ground like spaghetti noodles falling out of a big pot into a strainer. Carl didn't have a face anymore, but if he did it would have been buried in the dirt. Barry landed on his front with his face pointed to the left. Half of his jaw was missing and his eyes were open wide. His left hand fell to the ground last. It bounced slightly and then landed. The fingers were outstretched and slightly bent in an expression of horror. The ground under them was blood soaked and covered with entrails.

Larry decided not to go out the way he came in. He was fully charged and all of his relays were dried out. He aimed for the fence and flew forward as fast as he could. He ripped through the chain link fence like toilette paper. He dragged several strands behind him until they caught a tree and stayed there. The sound of the fence as it ripped was like a loud zipper and a bit like a spring. He leaped through the trees and bushes and moved through an open field. His wheels sank into the soft dirt and climbed out of it. They almost appeared to be floating. His front went up and down rhythmically. He looked like a speed boat on a large lake bouncing off of waves. Dirt flew up from his sides like water in the wake of a boat.

The noise brought people. Two customers were the first to get to the scene. One of them threw up as soon as he saw the bodies. In fifteen minutes police questioned everyone there and several policemen followed the tracks in the field.

Steve, Jackie, and George heard the broadcast on the police band. George found the address with Mapquest on his computer. In five minutes all three of them were out of the door and leaving the hotel parking lot. Steve followed George and George called Charlie on his cell phone at the stoplight. All three vehicles met at the junk yard. Steve and George climbed into the Hummer. Before Charlie could leave, the policeman from the other day flagged him down.

The policeman said, "Got room for one more?" Charlie said, "Hop in."

The policeman climbed in back. He pointed at the corner of the yard and said, "Go around the corner. He made a trail through that field." Charlie didn't ask, he just went where the policeman said. Once he found the trail, he followed in the same set of tracks. The tracks ended at a road and at an angle.

Steve said, "Don't follow those! He turned around and went the other way after he spun the mud off of his tires. He wants to hide near the scene. Go that way." Steve pointed in the other direction of the road. Charlie turned where Steve told him to. In the other direction of the road they saw police cars in the distance. One of the police cars turned around and followed the Hummer. It didn't turn on its lights or siren.

The policeman in the back seat talked with the squad car behind them and told them what they were doing. He spoke to Steve, "We have help. The car behind us is our back up."

Steve answered, "That's great unless he goes off road again. Then that squad car won't be able to follow."

The policeman said, "Let's hope he doesn't do that. Too late he did!" He pointed at a set of tracks running back toward the junk yard.

Steve said, "He sure did. Tell your friends to go to. Where could he be going?" Steve traced the roads on the map with his fingers to figure out where he was. He said, "Tell them to go here." He held his finger on a spot on the map and showed the policeman.

The policeman told the squad car behind them to go to the location Steve said. The squad car turned around and went the other way. The Hummer backed up a few feet and followed the SUV's tracks through another field. The ride was rough, but it was fast. The Hummer handled the terrain like it was born there. Everyone put their windows up because the dirt was flying into the passenger compartment. All passengers had seatbelts on. It was a wild ride. The front of the Hummer bounced hard up and down. It was a little bit like being on the high seas. The dirt was soft so they sailed through a rough ocean of dirt. Steve held his stomach because he was getting motion sick. He finally lowered his window, stuck his head out of the window and did what he had to do.

Steve said, "Sorry. I get seasick in rough water after a while."

Charlie said, "It's okay. I know you'll clean it later."

Steve stuck his head out again to do what he needed to and said, "Sure." His face was dark brown from the dirt.

George laughed and said, "You'd better look at yourself in the mirror."

Steve was confused, but did as George said. He looked at himself in the side view mirror and wiped his face with his shirt. In Steve's condition he had a little trouble speaking, "Can we catch up with this guy soon before I lose the rest of breakfast?"

George answered, "I'm right there with you buddy. I don't like it either."

Charlie said, "I think we are getting there soon. I see something and it is close to the junkyard." He pointed at the junkyard half a mile away to the right. A quarter mile ahead between the junkyard and a patch of trees was a road and several police cars. The lights weren't on, but the black and white colors were visible.

Steve said, "See if you can drive him out of there and to the police cars on the road ahead."

Charlie veered to the left and tried to herd the SUV to the police. He maneuvered so that he had the trees directly between him and the police on the road ahead. The police ahead figured out what the Hummer was doing with the help of the policeman in the back seat and his radio. Four off-road motorcycles took off into the sea of farm fields.

They were visible by the plumes of dirt they threw up in their wake.

Steve shouted, "Call them off now! The Hummer will kill them!" He pointed at the officer and sternly shouted, "Do it now! This SUV will eat them for breakfast. He is fully charged and weighs ten times as much as they do!"

The officer talked to his bosses in the department, "You'd better get those guys out of there. The suspect vehicle is too much for them."

Steve reached for the radio and took it from the policeman. He yelled, "Get your people out of there! They can't take him like this! He will roll over them like road kill. I saw him do it before! We are trying to deliver him to you!" The four brown clouds far away turned around and returned to the road.

Charlie brought the Hummer closer and a white streak blasted through the foliage. It moved like a streak of lightning until it hit the soft dirt, then it bounced up and down like a dog running through deep snow. As it went through the soft dirt it made a large cloud which almost hid the white paint. At first the brown cloud went toward the police on the road, and then it turned right toward the junk yard. Charlie turned right inside the SUV's turn. The Hummer was catching up. In just a short time the Hummer was about a hundred feet away from the SUV and the plumes behind the SUV made it very difficult to see through the windshield. Charlie backed off so he could see again.

In a few minutes, the Hummer was inside trees and then between rows of cars. Larry was just ahead and going through the rows of cars at the junkyard. He plowed through two squad cars at the gate of the junkyard. The Beast was waiting right outside of the junkyard and about twenty feet away from the squad cars.

Steve grabbed Charlie's arm and said, "Let me out here!"

Charlie stopped and Steve got out. George switched to the front seat and Steve climbed into the Beast. Jackie slid over.

Jackie was disappointed and said, "I wanted to do this."

Steve commented, "I'll bet you could, but this is my turn!"

Jackie grinned slightly at the off-handed compliment and said, "If I could, then why are you doing it?"

Steve said, "Because I really need this!" Steve goosed the accelerator hard and twisted his rear end around. In a second and a half he was pointed at the SUV and gaining ground. He passed the Hummer and got between the Hummer and SUV. The Beast got up to the tail of the SUV and bumped it hard.

Jackie said, "I guess you are taking this seriously!"

Steve said, "Oh, yeah! If I can stop him, then he won't kill anyone else!"

Jackie said, "So much for the body work!" Steve cried, "I'll get more!"

Steve pushed hard and leaned against the bumper of the SUV. The SUV slowed down to thirty miles per hour. The bumper of the SUV pushed hard against the fenders and grill of the Beast. The hood of the beast rose slightly and buckled, though it remained latched. The left headlight on the Beast shattered and a few pieces of broken glass flew over the windshield. Steve kept pushing.

The SUV sped up and they were heading toward traffic and buildings again. The vehicles separated. Steve stayed back because he was getting more damage than the SUV. The SUV veered left to the median and passed the traffic entering town like they were going backwards. The speed limit on this road was thirty five even though it opened up to four lanes. Steve stayed close so he could take advantage of any opportunities, but far enough away that he wouldn't hurt his vehicle any more than necessary. He was about fifty feet from the SUV and ready to veer left if necessary. The dirt here was covered with grass and firm. Steve had a bumpy ride, but he was able to keep up.

The SUV suddenly turned hard right and sliced across traffic through a very narrow hole. The vehicle in traffic which was nearest him slammed on its brakes and took a hit from a tailgater following too closely. Two more vehicles joined in the accident, but traffic in that lane finally slowed down and changed lanes.

Steve steered right but missed his chance to get across. He pushed his accelerator hard and tried to catch up to the hole which the SUV had used. He finally got across, but the SUV had opened the gap between them to several hundred feet. The Beast bounced up and down over the grass and occasional pieces of road. It was like riding a bull in a rodeo. His wheels were slow to steer because there was never any more than one wheel on the ground at any time. Steve slowed down so he could control his vehicle again. He needed to turn left to catch up with the SUV and he could only go straight while he was bouncing. Steve actually ended up almost facing oncoming traffic by the time he decided to slow down.

The Hummer was in traffic taking its time and caught up without any trouble. Charlie pulled off of the road and calmly called out to Steve, "Having fun?"

CHAPTER 20

STADIUM

Steve heard him, but turned left and regained pursuit of the SUV again. The SUV turned into a large parking lot. Ahead there was a small sports stadium. The SUV sailed through the fence and plowed through rows of seats as it descended toward the playing field. The seats were like theatre seats and set in concrete. The concrete was like a giant set of steps. Each time Larry hit a set of seats the seats collapsed forward and then shattered with pieces flying in many directions. From across the field, Larry's movement looked like a giant wave moving down the stands. The few people in this section panicked and fled to the right and left as Larry approached. The seats exploded forward at each row as Larry bounced down the large steps like a cat jumping down a set of stairs. Larry had enough forward momentum to keep from high centering and getting stuck. Steve stopped before he reached the fence. He told Jackie, "I can't do that!" He circled around the stadium and found a vehicle ramp down to the field.

The gates to the ramp were open and a car decorated with flowers and banners was slowly heading down it toward the field. Steve got closer and pushed the car ahead. He leaned on the gas pedal and the other car responded jerkily. The driver ahead shouted, "Are you nuts?" Steve pushed harder and both vehicles moved slightly faster down the ramp. The driver ahead pushed his brakes hard and Steve pushed him harder. The sound of tires rubbing against concrete screamed. The tunnel echoed the noise back like an amplifier. It was deafening. Finally the driver ahead accelerated out of the tunnel and pulled off to the left side. He yelled to Steve, "You're mine! I'm going to get you!" As soon as Steve passed, he turned and followed Steve. Steve turned right and headed toward the broken bleachers. Just then he noticed Larry turning toward him.

The stadium was full of people. A band was on the field at one end and teenagers in football uniforms were on each sideline. Larry was close to the middle of the field and aiming for the ramp which Steve had just come out of. In no time at all Larry was twenty feet from Steve. Steve turned sharply left, but misjudged the speed and turn. He missed and Larry slipped past. Larry missed the decorated car too and made his way up the ramp to leave the stadium. The car which Steve had pushed before was aimed directly at Steve's door and speeding up. The decorated car had a lot of extra weight because of a very high seat with someone on it and a wooden platform supporting a wildlife scene. Steve pushed on the gas hard and turned his wheels hard left. He took traction and pulled clear of the new attacker. The car which had been decorated as a float missed and stopped just short of hitting a couple of kids. The teenage girl on the seat above tumbled to the front seat head first. Steve was already gone.

Steve saw Larry leaving the parking lot and he followed. The Hummer and Oldsmobile were ahead of Steve. They had waited in the parking lot while Steve went to the field. Steve was catching up fast. Several squad cars arrived from the left. One of them tried to cut off Larry but Larry managed to turn inside and just bumped its rear end. That police car spun hard clockwise and ended up facing another oncoming squad car. Both cars collided. The sound was a low pitched thump which echoed.

Larry turned left down the road in front of the parking lot. He was headed out to the open country again. Three police cars in front of him took turns skidding sideways to block the road in front of him. Each time one of these cars stopped sideways in front of him, Larry turned hard to avoid it. One time he briefly left the road and drove through a yard. He got back onto the road again and had about a dozen vehicles in pursuit. Steve was in the middle of the group.

One of the policemen looked over at Steve and Jackie and sounded his siren. He said over his loud speaker, "Leave this area. This is police business. For your own safety, you are advised to slow down and pull over!"

Steve looked back at him and pointed at the SUV. He shook a fist at the SUV and yelled, "I am here to bring him down! I'm supposed to be here!"

That policeman shrugged his shoulders and shook his head. He thought then that Steve was just insane and decided not to waste any time with him. He turned his siren off and looked away.

Ahead of Steve the sound of gunshots echoed. One of the police cars got alongside Larry and tried to hit the hole with the globe in it. A few bullets got close to the globe and Larry turned hard right against the police car. The front of the police car hit Larry between the right doors and Larry twisted right. The white globe bounced out of its hole, bounced off of the front of the police car, and right back into its hole. The impact pushed the door inward and bent the frame slightly. Larry continued spinning in a clockwise direction until he stopped. He was facing the convoy of pursuers.

All of the cars stopped as fast as they could. There were no accidents this time. A cloud of dust rose as the wheels of a dozen vehicles skidded to a stop. There was enough dust that for a brief moment Larry disappeared in the dust. For about a minute there was silence. Nobody moved.

Car doors opened and many policemen climbed out of their cars. Their pistols were out in front of them so they could be ready for anything that the strange vehicle could do. Most of the policemen stayed behind their vehicles for safety, but a few ventured closer. By now almost all of the policemen were out of their cars.

Steve saw an opening in the cars and backed slowly out of the mess. He decided that being in front of the pistols was the wrong place for him to be. He also wanted to have room to move if he needed to. In less than a minute he was completely behind the group of vehicles.

Steve looked around. He saw miles and miles of miles and miles. Plowed under fields stretched out in every direction except the one from which he had come. The road they were on was large and wide. It was empty too. The shoulders were low. They dropped off to the fields below at a steep angle and went about five feet down. To the left was a large group of trees beyond the first field. Steve saw a road just past it.

Steve said to Jackie, "I wonder if he is going to make a break for it. There's a road over there. They can't follow him there either."

Jackie said, "I don't know, but how many of them are in their cars anymore?"

Steve said, "Hey wait a minute! He's charging so he can make a break for it!" Steve pulled forward and yelled to one of the officers, "Hey pull vehicles up against either side of that SUV. Box him in so he can't move. I just know he's going to take off across that field and go down that road! Don't let anyone approach that thing it will attack!"

The officer got onto his radio. His voice echoed through many squad cars. One of the officers at the other end of this parking lot got close to the Globe. He lowered his head so he could get a good look. He reached his hand out to touch the white ball and stroked the legs of it. He pulled his hand back quickly when he heard clicking inside it and saw the cameras on it move. He stood up and backed away in fear.

Larry suddenly sprang to life. He turned his wheels sharply to the right and twisted hard toward the side of the road. He accelerated quickly and jumped over the shoulder of the road. He didn't build up much speed in this short distance, so he landed clumsily in the dirt and bounced hard. His front bumper scraped the shoulder of the road, but his front wheels were so close to the bumper that they rolled down the incline. His wheels threw dirt everywhere and a cloud of dust covered the group of policemen. He moved as erratically as a puppet as he recovered from his risky maneuver.

A line of policemen formed against the edge of the road. They fired their weapons and reloaded to fire again. Since Larry had his front pointed at them he didn't receive much damage. The only target available to them which could really impair Larry was the globe on his roof. He got lucky. He bounced across the field like a rabbit. He sailed over the plowed furrows and left the large group behind. One of the squad cars tried to follow but got stuck at the bottom. It was wedged between the field and the rise of the shoulder with all four wheels in the air. The wheels were spinning furiously and the front wheels turned right and left. The vehicle remained in place. Finally the officers jumped out of their doors to the ground. One of them landed on his back. On the rise above, the other officers poked fun at this clumsy move.

Steve put the Beast in reverse and gunned it. He then put on the parking brake and twisted his wheels full left. He spun clockwise until he was pointed in the opposite direction and then goosed the accelerator.

Jackie knew what Steve was doing and looked on the map to figure out where they were. She held the map up and pointed with her fingers. She said, "We are here and that road is there. He is heading toward this."

Steve glanced over long enough to see the map and then turned back toward the road. He was accelerating as fast as he could. He knew that he could outrun the SUV. He just couldn't handle the farm fields. In his rearview mirror he saw the police cars beginning to move. They were following at a distance. They were faster than him but not ready. They were still getting into their cars. Steve saw a small dirt road veering off to the right and slowed down again so he could handle it. He drove down the small bumpy road at forty five and saw the SUV going down the other road. It was passing Steve.

Jackie laughed at it as she looked at it. She said, "It's all over the road. It is really messed up now."

Steve remarked, "That's good news. Let's hope that cut its top speed down a few more notches."

Jackie said, "Could we be that lucky?"

In minutes the entire convoy began to move in the same direction as Steve and the SUV. Steve was catching up to the SUV, but it veered right and left to keep Steve from coming alongside. The SUV was dog tracking terribly to the left. The frame was bent on the right side so the rear wheels were pointed slightly left. That caused it to move down the road somewhat sideways. The rear end was to the left of the front end. The SUV had to keep its front wheels a little left to go straight. Both vehicles were going about fifty miles per hour and heading back toward buildings and big roads. The convoy of followers began to gain ground and took the dirt road to get onto the road Steve and Larry were on. The Hummer and George's Olds were now at the back of the pack because they had been close to the front before.

Larry and Steve were still fighting it out on that road when the road opened to four lanes. Side streets and stoplights added cars everywhere around them. Both vehicles found themselves dodging slower cars. Larry managed to get several cars ahead of Steve at a stoplight He was in the right lane and so was Steve.

Larry stopped and waited at the stoplight because he had a barrier between himself and the Beast. The rest of the followers had not caught up yet either. He planned a move, but didn't want to forecast it to the rest of the group. Larry wanted to surprise them. He was worried about his reserves. He was now less than half charged because of all of the action. He wanted to escape fast and recharge as soon as possible. The gasoline engine and propane engine would help, but they could only run for a few minutes before they would overheat. He decided to risk it. He started both engines. The feeling of power was exhilarating. When the light turned green he leaped forward and took the lead of all of the other cars which had waited at the light. He then veered sharply left toward oncoming traffic. He just missed the car to his left and the oncoming cars and crossed the two oncoming lanes. He pulled onto a side road and turned right off of it. He had ducked behind buildings and was out of sight for a moment.

Steve was stuck between several cars and unable to move. He couldn't get to the left lane or median. He watched Larry disappear behind the buildings from the right lane and couldn't do anything about it. The rest of the group of pursuers couldn't do anything either. Steve stopped in traffic and waited for the other cars to pass. The sound of honking and rude remarks filled the air around him. Steve just shrugged his shoulders and pointed at the damage at the front of his car. Without saying a word, he told them that he was having car trouble. He didn't have a siren or anything else to get him through the traffic, so this had to do. The other vehicles in his group were also getting impatient too. Two police cars turned on their sirens and began to crawl through the traffic. Traffic in both directions got worse. In a moment Steve had room and was able to get to the median. He sat in the median and waited for traffic to clear so he could get to the other side. The two police cars with their sirens on backed up traffic in the other direction so Steve had to sit and wait for that to clear.

Larry knew he had a short time to make a decision so he thought quickly. He was behind a body shop and it was so tempting to hide in the empty stall. He looked around more. The body shop was to his right. Straight ahead was soft dirt leading to a five foot drop from a stone wall to a fairly busy street. To his left was a very quiet but run down neighborhood. These were very small single family houses and far apart. He saw abandoned vehicles everywhere. Many houses had boards across windows and furniture in the yard as if someone were throwing away everything in the houses. There were very few signs of life here. Larry risked the neighborhood.

Larry went forward into the soft dirt toward the road. He went all of the way forward until he felt the drop and then backed up. He then backed up in his own tracks so it would look like he went over the edge onto the street below. Larry proceeded to the neighborhood. He found a perfect spot three houses down. He found a house with a yard which had very little grass and open to the back yard on both sides. The sign in front said, "Sold to Brookridge Realties. Redevelopment project." Larry slowly and deliberately pulled into the yard. He looked behind with his rear view cameras to be sure that there were no tracks. As soon as he slipped between the houses to get to the back yard, he heard the noise of his pursuers. He made his way carefully to the back yard. He shut off both engines and remained perfectly quiet. The engines were not overheated, but warm. That meant that the antifreeze was up to temperature. At that temperature antifreeze boils at atmospheric pressure. Antifreeze whistled out of the radiator and squirted out onto the ground. The noise was uncomfortably loud for Larry and he quietly panicked. He looked around at the yard for a way out if he was discovered. There was a six foot wooden fence all around the back yard. He didn't know what was beyond the fence, so any escape would be a crap shoot. Larry hoped that he wouldn't be noticed despite the noise and if he was discovered, then he hoped that the yard next door didn't have an in ground swimming pool.

Larry's new yard had large trees which gave him cover overhead. It had very little grass, and it had an extension cord plugged into an outside socket just teasing him. The hissing noise from his radiator was quieter now so he wasn't worried anymore. He sat quietly and listened to the sounds of the police nearby. As he waited he charged slowly.

CHAPTER 21

FOUND

The police cars crowded the area behind the body shop where Larry left the tracks to the road below. A couple of squad cars moved randomly around the area to find tracks or signs of Larry. Steve arrived late to the scene and had to park on the road away from the body shop and tracks. He got out of the Beast and walked over to the most senior looking officer in sight. That officer was the one giving orders.

Steve got within five feet and said, "Sir, what do you have so far?"

The police officer acted instinctively and gave Steve short answers which sounded like orders, "I don't know who you are, but this is a crime scene and you don't belong here! My advice is to get back in your car and leave immediately before we decide you are obstructing justice!"

The detective Steve met on the night of Jose's death saw what was happening and walked up to the officer talking to Steve. He spoke softly, "Sergeant, this is the man who is helping us. He loaned us the software discs. You could say he is an expert."

The sergeant paused for a moment and decided how to rescue himself from embarrassment. He apologized and explained, "Sorry sir. We have tracks leading over the edge. We think the suspect vehicle went over the edge and joined traffic." He pointed at the tire tracks near the stone wall. He began again, "We want to be sure so we have a few patrols looking around the neighborhood. The next couple of blocks are deserted because of a redevelopment plan. This used to be a very bad neighborhood. We still chase squatters out from time to time."

Steve politely thanked the officer, "Thank you Sergeant.

Is it okay if I look at the tracks without touching them?" The Officer said, "Yes sir."

Steve walked over to the tracks and crouched down to look closer. He saw two tire tracks. The tracks were about six feet long. On each end of the tracks was a stone wall or paved parking surface. He stood up and walked to the stone wall to see if there were any more tracks. There were none. Steve began to piece it together in his mind. There would have been tracks in the grass near the other road. Five feet is not a big drop. There were only two sets of tire tracks. Larry was dog tracking. There should have been four sets of tracks.

Should he say anything? There were no forensic people here and he doubted that there would be. The police were just looking for something to follow; anything would do. The other cars pursuing Larry were probably too far behind to notice that he was moving sideways. Steve looked around. There were two units. The rest had gone, probably in both directions of the road ahead. Steve wrestled with the big question. Having a large group with him got in his way and kept him away from Larry. He could have gotten across the other street if those two squad cars had not stopped traffic and made it worse. Was it worth telling them what he thought?

Steve thought harder. His instincts told him that Larry was close enough to hear him breathe. The police didn't think the same way Steve did. No matter what Steve told them, they would assume that Larry used the tracks as a diversion and left them in the dust in another direction. Telling them wouldn't make any difference anyway. Steve decided to tell someone. The detective who had rescued him from the sergeant was still there. He was standing back and watching Steve think. Steve walked toward him.

The detective said, "Felling like a detective yet?"

Steve grinned and tried to be humble, "I don't think I'm ready for that yet. I would like to show you something though." He led the detective over to the tracks and pointed at them. He pointed out, "The SUV was hit on the right side and the frame was bent. It went down the road sideways. There should have been four sets of tracks." Steve pointed at the road below and said, "Is five feet high enough that he would have cleared the grass without bending a few blades of it? Do you think he went that way?"

The detective paused and looked at what Steve pointed out to him. He crouched down and looked at the tracks again. He stood up and looked in a straight line from the tracks and didn't see anything below. He turned away for a moment and thought about it. The detective then turned back toward Steve and patted him on the shoulder. He laughed and said, "Are you sure you don't want to be a detective?" The detective walked over to a squad car. The sergeant who talked to Steve before was inside. The detective pulled him over to the tracks and explained what Steve told him. The sergeant glanced at Steve and nodded in a gesture of respect. The sergeant rushed to his car and got on the radio. All Steve could hear was, "I need all units back here. He did not go down Morris Street!"

Steve turned around and walked away. The detective ran over and caught up with him. He asked, "Leaving already?"

Steve said, "With a hundred uniforms here I can't do much. I would just be in the way anyway. I would have stayed here to stake out the area, but you probably have that covered. I am going back to the hotel. I would like to hear from you once you know something."

The detective answered, "I will call you myself." He turned and went back to his car.

Steve got back to the Beast and sat down next to Jackie. He told her, "This place is going to be crowded soon. If I stick around I will probably be bored. Let's go back to the hotel."

Jackie said, "Pillow, sheets, HBO, pizza on the way there; okay let's go!" As they drove away it was starting to get dark. George and Charlie arrived over an hour later. The two men knocked on Steve's door. Steve got up out of bed, put on a housecoat and opened the door. He welcomed them in and offered them pizza. The four discussed the day and how they got separated. Steve did not get any calls that evening from the detective.

The next day, Steve and crew met in front of Steve's room. Steve told them what he thought, "I gave it some thought. I didn't hear anything from the police so they haven't found him yet." He laid a map out onto the hood of the Beast. He pointed to a spot on the map and said, "That was where the tracks were. We can be fairly sure that he was close to that area when I was there. He likes to hide close to the scene so he knows what the police are doing. It's a safe bet that once he found a hiding place he didn't move. The police patrolled the entire area. I'll bet they didn't go house to house. They probably looked in this area and past here."

George spoke up, "I see what you are saying, but why are you so sure that he stayed close to the area?"

Steve said, "When he took me for the ride, he circled the Barron plant. We found him a quarter mile from the Barron plant when we flushed him out. He shows tracks away, and then he turns around and gets close to the place nobody expects. He gets off on watching police look in the wrong place."

George said, "Kind of a wise guy."

Steve said, "Yes he is. I think he fears me because I am beginning to understand how he thinks. He only told me a bit more about what he thinks when he was sure that I could not escape from his passenger compartment. He can't hide from me as well as he can from the police because they think he put miles between himself and the last spot. I know he is right here." Steve pointed at the spot on the map where the tracks were. He said, "He is sitting there and laughing at us."

Charlie said, "Let's wipe that smile off of his face then!"

Steve said, "Thank you. That's what I had in mind. I want to hit him before he has a chance to recharge any more. George your car is very nice, but I don't think it would be a good idea to ram the SUV with it. I think you should ride with Charlie." George nodded. Steve looked at Jackie and said, "I would like for you to ride with Charlie too. I would feel better if you are in a large and safe vehicle." He looked at all of the faces around him and said, "I want to go house to house. You can cover me with the Hummer. I don't want anyone to get out of the vehicle while we do this. Once you are a pedestrian, you are dead meat. The Hummer is protection for all of you."

Jackie asked, "What about you?"

Steve walked around the driver side of the beast and said, "I have the ultimate getting even machine. The guy is so cocky that he doesn't think I can hurt him. It'll break his heart when an old relic wipes up the floor with his high tech masterpiece."

Jackie said, "He can still get to you."

Steve said, "I know he can, but I am more maneuverable than he is. My car is as light as a feather. It isn't much heavier than a Ford escort. It twists on a dime. I only need one good hit and he is finished. I can't lose." He looked at Jackie, "Besides, without you in the car I am 110 pounds lighter."

She said, "I weigh more than that"

Steve said, "Yeah, you have your purse. It weighs three pounds."

Jackie grinned at the compliment and climbed into the Hummer. George and Charlie got in too. Steve got into the Beast and everyone left the hotel parking lot. In a short time they arrived at the spot where Steve looked at the tracks. Steve got out and looked around at the area. The Body shop was closed and the overhead doors were closed. This was Sunday. He looked inside the windows and walked around to see if he could see anything. He did not see any sign of Larry at all.

George asked from the window of the Hummer, "Why did you look in there?"

Steve said, "If I were Larry, I would have come here. He needs body work."

George said, "Not at all. He is in great shape."

Steve laughed at the joke and said, "Great shape, no! Lethal, yes!" He was very serious when he spoke. He knew that as long as Larry could maneuver, no matter what his condition, he was about eight thousand pounds of terrible force. Steve climbed into his car. He drove into the yard of the first house, then drove far enough that he could look into the back yard and backed out. He drove across the yard of the second house and moved forward to the back yard and backed out again. He moved across the street and looked in three yards. The Hummer sat and waited on the street.

Steve pulled into the yard of the third house of the row he started on. He edged into the back yard. Something inside told him that this was dangerous. He pulled into the back yard far enough to see what was there. Steve was ready to move quickly and his right foot was itching to go through the floorboard. He caught sight of the SUV and lurched forward as hard as he could. His wheels grabbed the ground like they were glued to it and he felt the body of the Beast hurl itself forward. The Beast lost traction as he turned his front wheels left hard. His rear end almost lifted up as dirt flew in his wake and a large cloud of dust formed. He let up on the gas when he was almost pointed directly at the front wheels of the SUV. He thought, "Yes! I have the sweet spot now!"

Steve straightened his wheels and charged forward like a missile heading for its target. The cloud of dust covered the SUV. For the smallest moment he couldn't see the SUV at all, and then he saw it again. He was aimed at the rear bumper. He hit it hard, but he missed the rear wheels. He shattered the headlight on his right side and the pieces flew forward. They sprayed like a garden hose. The motion of the SUV combined with his own momentum twisted the Beast to the right and stopped it.

Larry was fully aware of Steve's presence as soon as he showed up at the body shop. He had waited until the last possible moment to move because he felt that it would be best if the two vehicles in pursuit were in different places.

He really wanted to deal with Steve one on one. The impact with the Beast twisted Larry toward the left, but the fence next door was a large target. He pushed hard with all four wheels and sliced through the fence like a hot knife through butter. Larry mowed over a 4x4 post like it wasn't there. He went through the fence slightly sideways. Splinters of wood and large chunks flew in all directions.

The yard next door was open except for a trampoline. He hit the trampoline and it lifted off of the ground. The frame of the trampoline bent and swayed as it lifted off of the ground. It unwillingly flew into the air and wobbled as it made its way to the roof of a nearby house. It looked a little bit like a flapjack rising up from a frying pan. There were no trees so Larry didn't have any real obstacles. He threw chunks of grass up in the air as he spun toward the left and aimed for the six foot wooden fence between the house next door and the limit of the side fence. He was feeling fine because he was about seventy five percent charged.

He erupted through the fence and brushed against the fence on the other side. He left stains of white paint on it and white flakes of paints became like powder and formed a small cloud behind Larry. The fence shook back and forth in his wake. He burst out between the house and fence at about twenty five miles per hour. The Hummer was sitting still against the curb waiting for Steve. It was to the right because Charlie was keeping a good position in case he needed to spring into action. Charlie began to roll a few seconds after he saw Larry. Larry turned sharply right and headed for a spot behind the Hummer.

Larry bounced over the curb, onto the street, and speeded up as he left the housing area. He bounced like a rocking horse once he landed on the paved street. Steve was in close pursuit. Steve hit the road a bit harder and almost scraped his front end as he recovered from the rough ride. Larry recovered a bit better than Steve and opened the distance between them. Charlie made a three point turn and followed behind Steve and Larry. Charlie trailed far behind, but still had Steve in sight. Jackie and George urged Charlie on to keep up with Steve.

Jackie said, "Move it now or I'll shoot!" She held up the crossbow so Charlie could see it in his rear view mirror. Charlie threw it into drive and goosed the accelerator. Larry had already disappeared around the corner but he still saw Steve turning right. Everyone's head went back when the hummer moved. It slipped a little bit on the loose dirt on top of the asphalt, but then the wheels grabbed hard and the front end rose slightly.

Larry was on a frontage road near the four lane road he was on the day before. On his right side was the housing area he had come out of. He took a hard right back into the housing area, and just as he took the turn, he saw the Hummer turning the corner to get on the frontage road. Steve was very close and just missed Larry's back bumper at the turn. Steve spun his rear wheels into the turn and bumped Larry as he came out of the turn. The road was home to the Beast. It just didn't like four wheel terrain. The Hummer was catching up quickly. Straight ahead was an elementary school.

CHAPTER 22

I'LL MEET YOU AT RECESS

The school was as abandoned as the rest of the neighborhood. There were boards in the windows and no cars in the parking lot. A large sign said "Brookridge Realty." Larry climbed out of the parking lot and went around the building. Steve skidded to a stop at the end of the parking lot. He turned the Beast so it approached the steep curb around the parking lot at an angle. He climbed over the curb one wheel at a time. Once he was onto the raised part of the school yard, he followed the SUV. Steve didn't mind destroying his vehicle by colliding with the SUV, but he didn't want to wastefully tear it up without inflicting harm on the SUV. The Hummer caught up with Steve in the parking lot and easily mounted the steep climb. Steve said, "Show off." He knew Charlie couldn't hear him, but he wanted to say it anyway.

Steve saw Larry on the playground by the adventure climbing set. Steve said, "I'm meeting you after school on the playground." Larry was stopped and pointed directly at

Steve. He was about 150 feet away from Steve and there was a clear passageway between them. The entire area had about two inches of sand on top of it. The rest of the sand was spread over the school grounds. This was the work of construction crews and many kids. Charlie stopped behind Steve and waited for a cue.

Jackie reached up to Charlie and grabbed his collar from the back seat. She yelled out, "Why aren't you doing something."

Charlie turned around and spoke, "This is Steve's turn. I'm only here if he needs me. If he were in here he'd say the same thing."

George said, "It looks to me like Larry has exactly the same thing in mind."

Jackie said, "So this is "High Noon? Are you insane? That thing over there has killed people and needs to be stopped. What do you want to do, play with it? This isn't some kind of stupid game. Wake up boys. Meeting on the playground during recess isn't what you should be doing!" George and Charlie were very quiet. Jackie sat back and mumbled, "You're a bunch of little boys."

Larry made the first move. He pointed directly at the front of the Beast. He wanted to take out the radiator and maybe the engine compartment. Steve knew instantly what Larry had in mind. Steve reacted quickly. He gunned his accelerator and turned his wheels right. The Beast twisted to the right. When Steve straightened his wheels and let up on the gas the car straightened out and grabbed the playground dirt. There was heavy playground equipment all around so

Larry could not turn inside and meet Steve. He had to go straight.

Larry barely nipped Steve's rear bumper. The impact was just enough to aim Steve at a large turtle made out of bars and metal plate. Its base was set in concrete so it wasn't going to move. Steve wasn't going very fast yet, so he didn't hit it hard enough to do much damage. His bumper hit a metal plate. His front left wheel then hit the plate. The entire front of the Beast lifted up about a foot. The front right wheel fell back down to the ground and the vehicle twisted to the right. It also pushed the Beast to the right. For a short moment the Beast was up on two wheels. If he had hit the turtle any harder he would have landed on his roof. Steve recovered quickly. He said, "Nice hit buddy." Steve punched the accelerator and got himself clear of Larry and anything else which could hurt him. He kept an eye on Larry as he moved to the other side of the playground. Both vehicles orbited the equipment counterclockwise looking for a chance to strike.

They were directly across from each other and Steve decided to make a move. He turned sharply and accelerated down an aisle directly toward the driver side of Larry. Steve was halfway down the aisle when Larry backed up hard and sat waiting for Steve. Steve slammed his brakes on and backed up. He knew immediately that Larry intended to hit him in the side and drive him into the monkey bars to Steve's left. Steve knew that this would disable his vehicle because it would get tangled in the pipes. Once Steve was out of action Larry could take on the Hummer and then come back to finish off Steve if he won.

Larry saw that Steve was wise to his plan and decided to try for a frontal hit. He had four inch pipe bracing his engine compartment. Steve did not. Larry turned hard and barreled down the aisle Steve was in and tried to ram the Beast in the front. Steve was looking behind so he wouldn't hit any equipment but he turned around when he heard dirt shift in front of him. He saw Larry gaining on him. Steve backed his vehicle as fast as he could and then turned hard right once he was clear of the equipment. The Beast twisted with the front end swinging wide. Steve put on the brakes when he was pointed at where he expected Larry to come out. He shifted it into drive and waited.

Larry didn't slow down like Steve expected him to. He exited the passageway through the equipment turning hard left. He was trying to ram Steve's front again. Steve had not gotten himself close enough to the equipment so Larry had room to turn. Steve learned his lesson very quickly. He turned right and aimed close to the opening. He wanted to hit Larry inside his turn. Steve's back end swung outside of the turn and made Steve almost parallel to Larry. Steve pushed as hard as he could. He bumped Larry's quarter panel. As soon as he felt the quarter panel, he turned his wheels left. He wasn't in contact with Larry for long because both vehicles slid and bounced off of each other. The scraping noise was enough to make Steve grit his teeth and it seemed like a boat bouncing off of a dock. The SUV weighed about three times as much as the Beast, so Steve felt more of the impact than Larry.

Steve cleared Larry and missed a piece of heavy play equipment. He turned right and headed around the perimeter of the playground again. He watched Larry as he did this. Larry hit a swing set because he couldn't stop in time after Steve pushed his rear end over to the right. Larry's front wheels were pointed left anyway, and when Steve wasn't there, he had nothing between him and the play equipment. Larry backed away from the swing set without much damage. The swing set was now very close to the ground because one of its legs was bent inward.

Steve knew how to play this game now. He watched Larry and planned his next move. He knew that Larry would win a head on collision. He also knew that just hitting Larry wasn't enough. He had to hit the right spot. Steve thought carefully and decided that hitting a front wheel at the trailing edge would bend the steering linkage and slow Larry down. He wouldn't be able to move very quickly and even though the Beast would be down hard, the Hummer could finish him off.

Steve was back at the big turtle that he bounced off of before. He saw Larry across the yard and there was a clear shot between them. He moved forward and presented his right side to Larry. He was still moving, but very slowly. He was trying to bait Larry. Steve was leaving the turtle on his right quarter panel. He was so close that his rear wheel rubbed against the bar holding it down on the ground.

Larry took the bait. He backed up and turned his wheels right. He stopped when he was aimed at the Beast. Larry charged the Beast. Sand and dirt flew up in his wake. He quickly picked up speed and approached Steve sideways because of his bent frame. A cloud formed behind him and Steve could hear the motors pushing the wheels. The sound was very quiet, but Steve was keenly aware of everything around him. Fear gave him more acute senses. Steve threw it into reverse directly from drive and pushed his pedal to the floor. He thought as he did this horrible thing to his transmission, "I have sinned!"

The Beast didn't move right away because it had trouble with the sand. His wheels spun furiously but didn't grab hold. Steve Pulled his foot off of the gas pedal and waited until he felt it grab hold. It inched backwards slowly. Larry was bearing down on him and picking up speed. He could hear Larry's wheel motor spinning faster. He could imagine what the impact would be like. He saw the passenger side of his car move toward him and the passenger seat end up in his lap. He was moving slowly in reverse now. He pushed a little harder on the accelerator and the car backed faster. Larry turned left to meet Steve. Steve pushed a little harder on his gas pedal and moved behind the turtle. Now Steve was moving quickly again.

Larry turned left again to get to Steve. The turtle was directly between Steve and Larry now. Larry figured that he would land on top of Steve or high on the passenger compartment. If he missed he could land on the hood which would ruin the engine compartment. It he stopped the Beast then he could come back and finish off Steve after he took care of the Hummer. He continued to pick up speed and approached the turtle.

Steve punched the gas right before Larry got to the turtle. The Beast's rear wheels threw dirt high up in the air and a dust cloud covered the Beast completely. He felt his wheels grab the dirt and pull him backward. He was picking up speed. He kept moving until he heard the wheels of the SUV go over the turtle and then land in the dirt to his left.

Steve released the pedal to clear the cloud of dust. He saw Larry heading away to the left. He turned his wheels right and ducked back into the jungle of play equipment. He began to do a zigzag pattern through the maze of steel bars and plates. He kept his attention fixed on Larry as he did this.

Larry recovered from the jump over the turtle and returned to circle the playground equipment in a clockwise direction. Steve ended up under the adventure play set where he first saw Larry. Steve backed between the bars and to the spot where Larry had been before. He was pointed directly at Larry's right side. He could see the white dome tucked into the side. Larry backed up and turned to face Steve. He began to advance. Steve sat there and waited.

Steve's heart pounded loudly in his ears. He shook with fear. He planned his next move as he watched the menacing vehicle approach. The SUV was almost halfway down the narrow aisle and picking up speed. Steve backed up and turned his front wheels left. He moved slowly at first, but carefully pushed harder on the gas and picked up speed.

He put the selector in drive and goosed the accelerator. He slid against the adventure play set and sat there waiting for Larry to pass.

Larry watched Steve do this and almost laughed at him. Larry knew that he had the advantage of weight and size. He also counted heavily on the four inch pipe supporting his engine compartment. He decided to speed up and see if Steve could hit him at all. He wasn't worried. Larry decided that Steve was likely to run like he had twice already and even if he did try something, Steve couldn't hurt this big of a vehicle very much. Larry said, "You're mine now!"

Steve backed up about ten feet and began to move forward. He pushed as hard as he could against the sand and dirt without losing traction. When Larry arrived he met him at the front edge of Larry's right fender. Steve pushed him into a large set of steel bars. The forward momentum of both vehicles shoved Larry's left side against the bars and twisted Steve to the right. Steve backed up while he had Larry stopped and rammed his target area as hard as he could. He hit the right front tire from behind and felt it give way. Steve backed away again. A green cloud rose from the front of his vehicle and the sweet smell of antifreeze nauseated him. Steve knew that he only had one more chance because the Beast had just lost its radiator. He backed up, hit the wheel again, and shut his car off. The blood soaked right fender of the SUV got an antifreeze bath from Steve's car. The antifreeze sprayed out onto Larry's fender and made a loud whistling sound.

Steve pushed his door open. He had to work at it. The fender was lodged tightly against the door because of the damage it suffered. When he opened his door he shoved the pointed forward edge of his door into the fender and bent it inward. He shook his head as he did this because he really cared about his car. Everything was quiet now and the SUV didn't move at all. The only sound was the Beast's radiator and the contracting of the metal cooling in the engine of the Beast. It was a popping and snapping sound.

Steve stood next to the Beast and stroked the roof. He said quietly, "You did good girl." He looked around to see what happened. He looked at the wheel on the SUV. It was turned right. The rim was cracked on its trailing edge. Three out of five spokes were broken. The route behind the SUV was partially blocked because of bent poles of playground equipment. He went around behind the Beast and the front of the SUV to see which way its other wheel was pointing. He had to be sure.

Once he was in front of the SUV he heard something. There was a quiet clicking sound like computer thinking and a scraping sound. He ran past the front of the vehicle to the other side and saw the other wheel pointed left. He was alarmed because of the noises and headed for cover behind the Beast. Before he got very far from the front of the SUV it moved toward him. It had great trouble moving because its front wheels were pointed in opposite directions, but it quickly corrected for this problem by using three wheels to move and dragging the fourth.

The SUV shuddered violently and rocked randomly from side to side. It moved very slowly now. Dirt flew up in all directions as it moved. It also had to push the Beast out of the way because the Beast was wedged against its right wheel. Steve backed slowly away as eleven thousand pounds of steel inched toward him. The Beast rocked and bounced as its tires were lifted out of the holes they dug when Steve rammed the wheel.

Steve shouted to Larry, "Do you need some help moving Buddy?"

Larry yelled in frustration over his radio, "You're going to need some help getting to your car when I catch you. Your car's going to need some help too."

Steve replied, "That's okay! I can fix it! Who's going to fix you?" The pile of metal continued moving toward Steve. The Beast lifted up and down as it was pushed by Larry. Bending metal whines and scraping noises filled the air. The Beast was now about four feet away from Steve and Larry was almost clear of it. Larry lurched forward after he cleared the Beast. He charged at Steve at a speed which almost rivaled a fast walk. He bounced right and left as he dragged his left tire over hills of sand.

Steve teased the monster. He pranced around in front of it and drummed on the hood with his hands. He yelled, "We can play this game all day! How are your batteries doing in this sand? I used a lot of gas messing with you!"

Larry replied, "See that bright thing up there? It charges me. It's nice of you to worry so."

Steve had led Larry about thirty feet away from the Beast. He thought he could keep taunting Larry enough to wear him out, but Larry tried something new. Larry turned his wheels in reverse and bounced backwards at about ten miles per hour. He still bounced because of the bent steering linkage, but he had enough control to maneuver. He aimed for the tail of the Beast and rammed its rear end. The Beast lifted up from the impact and the rear was crushed. The trunk flew wide open. The back tire of the Beast was bent inward. The Beast landed about five feet back from Larry. Larry twisted hard counterclockwise and aimed his rear end at Steve.

Steve wasn't so smug anymore. He made the same mistake Larry had and got too cocky. Steve thought quickly. He figured that Larry could turn left better than he could right, so he dodged right. Larry still caught him, but just barely. Steve landed on the ground and rolled to the right before Larry could run over him. Steve got up fast and ran to the front of Larry where he had the advantage. He looked around to see what he could use. He saw the adventure play set. It was in front of Larry, so Steve had at least a little time to get there. Steve ran for it and began to climb. Once he was about eight feet above the ground he looked down at Larry and yelled out, "Missed me again. You must be getting old!"

The adventure play set was very solidly constructed. The posts were as big as telephone poles and set deep in the ground. Steve figured that Larry could go through these, but not very easily. Steve was up in the pirate ship part and had a helm in front of him. He looked down and saw that the open trunk of the Beast was about five feet away from the base of the platform he was on.

Jackie shook Charlie's collar and said, "Will you do something now?"

Charlie put his gear selector in drive and orbited the equipment to get to Steve so he could rescue him from his perch.

Steve saw Charlie and held his hand up toward him in a "Stop" gesture. Steve looked back at Larry and then back at the Hummer and shook his head at the Hummer in a gesture of, "No!" He then waved the Hummer away.

Jackie shook Charlie's collar and yelled, "Don't listen to him! Go over there!"

Charlie put on his brakes and turned around to look at Jackie. He said, "I don't know what he has in mind, but if I go over there I will probably mess it up. Let the man do his thing!"

Jackie yelled, "He doesn't have a thing you coward! He is stuck on a kiddy tower and acting like he can do something. Save him now!"

Charlie spoke up, "I am not a coward! I don't know anything about his thing! If you don't stop now I will back behind this building and you won't be able to see anything! Are we clear?"

Jackie mumbled, "Yes we are perfectly clear." She crossed her arms and leaned back in the back seat with an angry look on her face. She thought quietly, "Men and their stupid playground rules!"

Steve was glad that the Hummer backed off. He didn't want them in danger because he just thought of a plan. He had told the "Rain Man" before that he wanted to pull a road flare out of his trunk and incinerate him. Steve thought how nicely the interior of the SUV would burn. Steve jumped out onto the wobbly bridge of the play set and yelled to Larry, "Hey stupid!"

Larry charged for a short distance, and then Steve leaped onto the next platform.

Steve yelled down, "Your Momma was a tricycle!"

Larry edged closer to the second tower. He yelled back, "Do you want to know what I did with your Momma?"

Steve answered, "Maybe later!" He jumped across the bridge in two steps and leaped out of the first tower. He landed awkwardly on the ground and rolled back up to his feet. He missed the trunk of the Beast and ran back to it. He found the road flares and pulled one out of the emergency road side kit. He held it in his right hand and ran toward the front of the SUV which was now pointed at him. Steve was now about five feet from the front of the SUV and it began to back away.

Larry tried desperately to turn around so he could aim his rear at Steve. With Steve's running start he caught up with the SUV and jumped aboard. He picked a spot on the SUV's hood. He stuck his right leg into the hole where the windshield used to be to hold himself steady. He rested his foot on the driver seat. Larry bucked like a bull as Steve tried to light the road flare. Steve swayed back and forth, but pulled the cap off of the flare and rubbed it like a match against the reactant side. The flare started instantly and the light blinded Steve. He couldn't see what he was doing for a moment, so he just dropped it into the hole by his leg. Once Steve knew that the flare was inside the passenger compartment of the SUV, he lifted his leg out of the hole and jumped away from the direction Larry was moving. He still could not see very well, but he rolled on the ground and got up.

Steve was still blind from the road flare. He could make out some shapes and listened carefully for Larry. Steve had not looked at the flare for very long so he recovered quickly. He saw the back end of the SUV coming at him and he dodged it in the direction of the adventure play set. This time Larry missed him. Steve reached the play set and climbed up to the first tower to watch. Steve caught his breath and watched Larry.

Larry was moving randomly and away. Smoke billowed out of the places where windows used to be. His wheels slowed down and he only crawled slowly through the sand. Eventually Larry didn't move at all. Flames poured out of the passenger compartment of Larry and began to climb higher. The sound of flames crackling became louder.

Steve climbed out of the tower and motioned the Hummer closer. He was the first one to get to the other side. Steve wanted to know that the brain had shut down. When he got there he saw that the white globe was now brown and using its legs to climb down a wire bundle to the ground. It reached the ground and used its legs to unhook the cable to the SUV. It looked like a large spider as it went down the wires and tried to run away along the ground. The spider didn't move very well in the sand and Steve reached it with no trouble at all.

Steve picked it up like a basketball with his hands and burned his fingers. The arms of the ball tried to reach Steve's hands, but they weren't long enough. They waved in the air like a spider trying to catch a grasshopper. Steve turned the thing upside down and planted it in the dirt that way. The arms waved back and forth trying to right the globe. Steve got up and ran back to the trunk of the Beast and returned with a pair of wire cutters. He cut the legs short one by one until six two inch stumps waved around in the air. He picked up the globe again and carried it further away from the burning SUV. He laid it on the ground and the Hummer stopped about fifteen feet from him.

Steve watched Jackie, George, and Charlie get out of the Hummer. He glanced at the destruction behind him and thought about what was near the fire. He thought out loud so his friends could hear him, "Gas tank; propane tank; batteries. They are going to blow! Let's back away!" Steve stood up and ran past the Hummer and his three friends followed him. Steve stopped about fifteen feet past the Hummer and said, I think we are safe here. He stood up watching the SUV.

The smoke leaving the SUV became black instead of gray. Yellow flames licked out of the holes everywhere like snake tongues. Suddenly there was an ear splitting explosion. The SUV suddenly fell at the right front end. A tire had exploded. Flames shot out of the front end of the SUV like exhaust from a jet engine. The sound was a loud and low pitched, "Whoosh!" Another explosion followed. The front of the SUV dropped again. The other front tire blew.

Jackie walked over to Steve and said, "This was your thing?"

Steve said, "Huh?"

Jackie explained, "Charlie said that you were doing your thing when you were up on the tower telling us to back away."

Steve looked at Charlie and George and replied, "Yes it was my thing. It's just one of those things. I did the tower thing and flare thing. I didn't like the bouncing SUV thing much. The falling thing hurt a bit. Was it okay for you?"

Jackie laughed and said, "The falling thing was kind of funny. Let's do it again. This time I need a camera so I can send it off to a TV station."

Steve replied, "Okay, I'll put out the fire and plug this back in." He walked over and picked up the brown globe.

Jackie said, "That's okay. Your thing was just fine the way it was."

Steve carried the brown globe over to her and the gas tank exploded. The entire rear end of the SUV lifted up with the explosion. Dust flew in all directions and a large metal piece flew past Steve. He heard the sound of it moving past his ears. Steve held Jackie's arm and led her further away from the SUV. He said, "There's one thing I haven't heard go off yet."

Jackie asked, "What?"

Steve said, "If that was the gas tank, then it would be the propane tank. If it's the other way around, then you get the picture."

George and Charlie caught up with Steve. Charlie asked, "What about the Hummer?"

Steve looked back and said, "It's far enough away and it can take more than we can. It's not hurt yet is it?"

Charlie looked back and said, "Not yet."

Another large explosion lifted the SUV and bent the roof straight again. Pieces of the SUV rained all over the Hummer. Some of the pieces were still on fire. Charlie looked back and said, "Now it is."

Jackie and George stood still as Steve and Charlie ran back to the Hummer. Steve yelled out, "That should be the last big explosion!"

As Charlie and Steve removed burning chunks of SUV from the roof and hood of the Hummer there was another small explosion. Charlie looked at Steve and Steve said, "Just another tire." In two minutes the debris was off of the Hummer and they both got in. Charlie started it up and backed it to Jackie and George. He shut it off and they both got out. Steve got out and walked up to George he said, "Hey George would you call the fire department and a tow truck?"

George said, "Sure, but you want to tow that monster?" Steve pointed at the Beast and said, "No, that one."

George pulled out his cell phone and made the call. He walked away as he spoke to the other party.

Charlie said, "Maybe you should tow that one too. How much would Barron pay for it?" Charlie stared at the burning SUV for a moment and said, "I don't think I'll get my arrows back."

Far away in a hidden underground facility a quarter mile beneath the ground there was a large computer complex. A room as big as a gymnasium, had computer banks against one side. The other side was lined with computer stations. Men and women in white lab coats stood in a corner of a room and listened to their instructions from a black man in a black suit. The man said, "I am agent Brian Stills." He held up a CD-RW disc for everyone to see and said, "I want to talk to this guy!"

www.ingramcontent.com/pod-product-compliance
Lightning Source LLC
Chambersburg PA
CBHW071241300726
48975CB00002B/508